BIBLIOGRAPHY OF CONGRESSIONAL GEOLOGY

BY

HAROLD R. PESTANA

Department of Geology

Colby College

HAFNER PUBLISHING COMPANY
New York
1972

Copyright © 1972, by Hafner Publishing Company, Inc.

Published by

Hafner Publishing Company, Inc.
866 Third Avenue
New York, N.Y. 10022

Library of Congress Catalog Card Number: 79-175939

A
557.3
P476b
1972

All rights reserved. No part of this book may be reproduced or transmitted in any form or by any means, electronic or mechanical, including photocopying, recording or by any information storage and retrieval system, without permission in writing from the Publisher.

Printed in U.S.A. by
NOBLE OFFSET PRINTERS, INC.
NEW YORK 3, N. Y.

Part 1

INTRODUCTION

This bibliography includes and indexes all of the geologic documents published from 1818 to 1907 in the *Congressional Documents Set*. This set of documents began with the 1st Session of the 15th Congress in 1818 and continues to the present day. From 1818 and through the 2nd Session of the 59th Congress in 1907, the *Congressional Documents Set* contains many types of geologic reports. Starting with the 60th Congress, publication and distribution of scientific materials as Congressional documents was greatly curtailed.

The *Congressional Documents Set* consists of House and Senate committee reports on legislation and other documents of legislative or executive origin. Many of these documents contain significant, or historically important geologic material. The set contains several series of documents. For the 15th Congress there are two series, *House Documents* and *Senate Documents*. From the 16th through the 29th Congress, an additional series, *House Reports,* was added. During this time, reports of Senate Committees were published as *Senate Documents.* From the 30th through the 53rd Congress six different series were present: *Senate Executive Documents, Senate Miscellaneous Documents, Senate Reports, House Executive Documents, House Miscellaneous Documents,* and *House Reports.* In these series the executive documents are usually reports from the executive branch to Congress, such as annual reports of executive agencies and reports of explorations and surveys. The miscellaneous documents consist of publications that Congress voted to include. Starting with the 54th Congress, there was a reduction to four series of documents: *Senate Documents, Senate Reports, House Documents,* and *House Reports.*

Geologic material found in the Congressional Documents Set

The geologic documents in the *Congressional Documents Set* deal with nearly all aspects of the science and are published through a number of government branches or agencies. A brief description of these various documents is found in Pestana and Bonta (1970). Especially well represented are documents dealing with geologic exploration and documents important to the history of geology. Publications of the U. S. Geological Survey and documents relating to the origin of the Survey are found in the set. The Smithsonian Institution, The Coast Survey, The Army Corps of Engineers, The Bureau of Soils, The National Academy of Sciences, The Treasury Department, and The Director of the Mint are some of the sources of the geologic documents found in the *Congressional Documents Set.*

This set is an unrecognized source of geologic literature. Unrecognized because most of the geologic documents of the *Congressional Documents Set* are not listed in the usual sources. Nickles' great work, *Geologic Literature on North America 1785-1918* (1924) (usually called the *Bibliography of North American Geology*) is deficient in several respects. First, it is restricted to North America; second, it does not include all geologic articles; and third, it often fails to indicate that a publication may *also* be found in the *Congressional Documents Set.* A preliminary study of this set in 1967-68, indicated that 40% of the geologic documents are not listed in Nickles (1924) and that an additional 30% are not noted as also occurring in the *Congressional Documents Set.* Other catalogs and indexes such as Poore (1885), Ames (1905), and Hartwell (1911), are either incomplete or difficult to use in locating geologic reports. The great 3 volume bibliography, "The Pioneer Century", (Meisel, 1924, 1926, & 1929) includes geologic documents found in the *Congressional Documents Set* up through 1865, but it does have omissions and entry is only by author and expedition. Meisel, however, is very useful in listing the table of contents of many of the larger government publications.

Distribution of Congressional Documents

During the early years of the *Congressional Documents Set* distribution of documents was somewhat random and unreliable. Efforts were made to insure a continuous supply of documents to libraries, but it was not until 1859 that legislation established depository libraries and made distribution consistant (Boyd and Rips, 1949). All of the *Congressional Documents Set* is available on microfilm, and a list of depository libraries is given in Schmeckebier and Eastin (p. 445-456, 1961).

Introduction 3

In 1895 the volumes of the *Congressional Documents Set* were serially numbered. Over 3400 volumes were numbered retroactively, and the serial numbering scheme has continued to the present.

Inclusions in this Bibliography

Because geology is related to and merges with many other disciplines, the problem of what and how much to include was constantly present during the compilation of this bibliography. An attempt was made to include all publications that would be included in a current bibliography of geology. The following will give an idea of what has been included and what has been excluded with respect to commonly encountered overlapping areas.

Geology/Oceanography: Documents dealing with ocean bottom features, shoreline changes, sediment description, and methods of sampling sediments have been included. Documents dealing with currents, water temperatures, or tidal motions have been excluded.

Geology/Geophysics: Included are publications dealing with earth temperatures, electricity, gravity, or magnetism. Excluded are those dealing with the precise determination of position by magnetism, and those dealing with triangulation and mapping by the Coast Survey.

Geology/Exploration: If a journal or report of exploration includes routine mention of lithology, structure, or topographic description it has been included. If a specific part of a journal or report is devoted to geology, this part in included. Excluded are documents dealing exclusively with land surveying, and journals or reports that contain no mention of geologic data. These latter generally consist of descriptions of indians seen, where camp was made, miles travelled per day, etc.

Geology/Hydraulic Engineering: Documents containing significant geologic information such as sediment description, channel geometry, topographic description, or drainage patterns are included. Those that deal primarily with surveying and improvement of navigation are excluded.

Geology/Road or Railroad Construction: Included are reports that contain lithologic and topographic description and those that describe geologic exploration. Excluded are those that only describe the surveying or the progress of surveying.

Geology/Mining: Included are publications covering lithologic or structural description, origin of ores, or stratigraphy. Documents on mining law and mining schools are included when these occur in a larger work on Economic Geology. Excluded are documents dealing with legality of claims, production statistics of mines, and mine safety.

U. S. Geological Survey Documents

Because the U. S. Geological Survey publications are listed in the usual bibliographies, they are not included in the main body of this bibliography. However, because the *Congressional Documents Set* is a good source for early survey documents, a complete list of these and their location is given in part 2.

Abbreviations and Annotations

The following abreviations are used in this bibliography: HD for *House Document,* SD for *Senate Document,* HED for *House Executive Document,* SED for *Senate Executive Document,* HMD for *House Miscellaneous Document,* SMD for *Senate Miscellaneous Document,* HR for *House Report,* and SR for *Senate Report.* Following the abbreviation of the document series is the number of the Document, the volume number of the series, the number of the Congress and Session, the pagination, and finally the serial number in parentheses. For example:
Cuvier, Georges, 1861, Memoir of Rene-Just Haüy: SMD 21, v. 1, 36-2, p. 376-392, (1089).
This article is in Senate Miscellaneous Document 21, which is in volume 1 of the Senate Miscellaneous Documents published by the 2nd Session of the 36th Congress. This document is found in serial volume 1089 of the *Congressional Documents Set.*

The brief annotation found with most inclusions will a) indicate other occurrences in the *Congressional Documents Set,* or b) indicate if it is part of a larger document that is also available elsewhere i.e. Annual Reports of the Smithsonian Institution, or c) describe the geologic content to a greater extent than the title. Annotations have not been made for documents whose titles aptly describe the contents, such as the reports of soil surveys of various counties.

The Index

The index of this bibliography has been compiled by using as a model the "Guide to indexing bibliographies and abstract journals of the U. S. Geological Survey". The index includes as many cross-references as feasable and, except for a few minor changes to suit the material, is the same as that in the current *Bibliography of North American Geology*. A few broad area terms, such as *Great Plains* and *Great Basin*, and a few general terms such as *Geology, History of-, Geology, Methods-,* and *Geology, Relation to other sciences-,* have been added. The term *Memoir* has been used instead of Biography, because most biographical sketches are so titled in the *Congressional Documents Set.*

Acknowledgements

I wish to thank the following who have helped in various ways in the preparation of this bibliography. Mr. George E. Becraft, Chief of the Office of Technical Reports of the U. S. G. S., for supplying a copy of the indexing guide used by the Survey. Mr. Bruce D. Bonta, Readers Services Librarian, Colby College, for introducing me to the *Congressional Documents Set,* and for expending a great deal of effort and time in advising me during various stages of this project. Mr. Robert E. French for locating the articles published from the 56th Congress through the 59th Congress and for the extra effort of travelling to other libraries to check volumes missing from the Colby College copy of the *Congressional Documents Set.* Dr. George W. White, Research Professor of Geology, University of Illinois and Vice President on the International Committee for the History of the Geological Sciences, for suggestions and a great deal of help concerning the manuscript. The Committee on Research, Travel, and Sabbaticals, of Colby College for financial assistance in preparing the manuscript. Mrs. Doris L. Downing for the difficult task of typing the index from the original cards, and Miss Amy Becker for help in reviewing the manuscript.

References Cited

Ames, John G., 1905, Comprehensive index to the publications of the United States Government 1881-1893: HD 754, v. 119-120, 58-2, 1590 p., (4745-4746).

Boyd, Anne M., and Rips, Rae E., 1949, United States government publications: New York, H. W. Wilson Co., 627 p.

Hartwell, M. A., *Editor,* 1911, Checklist of United States public documents 1789-1909: Washington, Government Printing Office, 1707 p.

Meisel, Max, 1924, 1926, 1929, A bibliography of American natural history: the pioneer century, 1769-1865, 3 vols., Premier Publishing Co., Brooklyn, New York.

Nickles, John M., 1924, Geologic literature on North America: U. S. Geological Survey Bulls. 746-747, 1167 p. and 658 p.

Pestana, Harold R., and Bonta, Bruce D., 1970, Congressional Geology: Geol. Soc. America Bull., v. 81, p. 899-904.

Poore, Ben P., 1885, A descriptive catalogue of the government publications of the United States, September 5, 1744–March 4, 1881: Washington, Government Printing Office, 1392 p.

Schmeckebier, Lawrence F., and Eastin, Roy B., 1961, Government publications and their use: Washington, Bookings Institute, 476 p.

Part 2
UNITED STATES GEOLOGICAL SURVEY PUBLICATIONS FOUND IN THE CONGRESSIONAL DOCUMENTS SET

The following documents are excluded from the main body of this bibliography because they are well known and are listed in the usual sources such as the *Bibliography of North American Geology*. They are included here because the *Congressional Documents Set* is a generally unrecognized source of these out of print publications. After the 2nd Session of the 59th Congress in 1907, distribution of U. S. Geological Survey publications *as congressional documents* was restricated. Although assigned document numbers they were not distributed to depository libraries as part of the *Congressional Documents Set*. One exception appears to be the *Water Supply Papers* which were distributed up through number 320, published by the 3rd Session of the 62nd Congress in 1913. Abbreviations used in the following list are the same as those in the main body of the bibliography and are explained in the introduction.

ANNUAL REPORTS OF THE U. S. GEOLOGICAL SURVEY
1   HED 1, part 5, v. 10, 46-3, (1960).
2   HED 1, part 5, v. 11, 47-1, (2019).
3   HED 1, part 5, v. 12, 47-2, (2101).
4   HED 1, part 5, v. 12, 48-1, (2192).
5   HED 1, part 5, v. 13, 48-2, (2288).
6   HED 1, part 5, v. 13, 49-1, (2380).
7   HED 1, part 5, v. 10, 49-2, (2469).
8   HED 1, part 5, v. 12, part 1-2, 50-1, (2543-2544).
9   HED 1, part 5, v. 13, 50-2, (2639).
10   HED 1, part 5, v. 14, part 1-2, 51-1, (2727-2728).
11   HED 1, part 5, v. 14-15, 51-2, (2843-2844).
12   HED 1, part 5, v. 17-18, 52-1, (2936-2937).
13   HED 1, part 5, v. 15-17, 52-2, (3090-3092).
14   HED 1, part 5, v. 16-17, 53-2, (3212-3213).

15	HED 1, part 5, v. 17, 53-3, (3308).
16	HD 5, v. 17-20, 54-1, (3384-3387).
17	HD 5, v. 15-17, part 2, 54-2, (3491-3493.2).
18	HD 5, v. 15-20, 55-2, (3643-3648).
19	HD 5, v. 17-23, 55-3, (3759-3765).
20	HD 5, v. 22-29, 56-1, (3919-3926).
21	HD 5, v. 31-39, 56-2, (4105-4113).
22	HD 5, v. 28-31, 57-1, (4295-4298).
23	HD 5, v. 24, 57-2, (4463).
24	HD 5, v. 24, 58-2, (4650).
25	HD 5, v. 24, 58-3, (4803).
26	HD 5, v. 22, 59-1, (4962).
27	HD 5, v. 18, 59-2, (5121).

BULLETINS OF THE U. S. GEOLOGICAL SURVEY

1	HMD 16, v. 1, 47-2, (2115).
1-6	HMD 71, v. 36, 48-1, (2248).
7-14	HMD 41, v. 16, 48-2, (2326).
15-23	HMD 33, v. 1, 49-1, (2406).
24-36	HMD 163-164, v. 8, 49-2, (2495).
37-41	HMD 375, v. 2, 50-1, (2566).
42-46	HMD 137, v. 11, 50-2, (2664.1).
47-54	HMD 138, v. 11, 50-2, (2664.2).
55-61	HMD 244, v. 32, 51-1, (2791).
62-65	HMD 136, v. 15, 51-2, (2883).
66-71	HMD 21-22, v. 17, 52-1, (2975).
72-75	HMD 23, v. 18, 52-1, (2976).
76-80	HMD 24, v. 19, 52-1, (2977).
81-83	HMD 25 & 336, v. 20, 52-1, (2978).
84-86	HMD 337, v. 45, 52-1, (3003).
87-89	HMD 176, v. 23, 53-2, (3251).
90-97	HMD 177, v. 24, 53-2, (3252).
98-101	HMD 178, v. 25, 53-2, (3253).
102-106	HMD 179, v. 26, 53-2, (3254).
107-117	HMD 180, v. 27, 53-2, (3255).
118-122	HMD 78, v. 9, 53-3, (3335).
123-126, 128-129	HD 311-314, 413-414, v. 77, 54-1, (3444).
127	HD 412, v. 78, 54-1, (3445).
130-136	HD 415-421, v. 79, 54-1, (3446).
137-142	HD 27-32, v. 35, 54-2, (3511).
143-149	HD 33-37, 340, 347, v. 36, 54-2, (3512).
150-152	HD 567-569, v. 74, 55-2, (3702).
153-156	HD 570-572, 574, v. 75, 55-2, (3703).
157-159	HD 302-304, v. 87, 55-3, (3829).
160-161	HD 305-306, v. 88, 55-3, (3830).
162-169	HD 17, 572, 633, 714-718, v. 54, 56-1, (3951).
170-176	HD 719-720, 746-750, v. 55, 65-1, (3952).

177-178	HD 535-536, v. 113, 56-2, (4187).
179	HD 461, v. 97, 57-1, (4364).
180-184	HD 462-466, v. 98, 57-1, (4365).
185-187	HD 467-469, v. 99, 57-1, (4366).
188-189	HD 470-471, v. 100, 57-1, (4367).
190-193	HD 472-475, v. 101, 57-1, (4368).
194-203	HD 635, 693-701, v. 102, 57-1, (4369).
204-209	HD 204-208, 430, v. 65, 57-2, (4504).
210-213	HD 434-437, v. 66, 57-2, (4505).
214-216	HD 470-472, v. 67, 57-2, (4506).
217-222	HD 62-66, 606, v. 58, 58-2, (4648).
223-225	HD 675-677, v. 59, 58-2, (4685).
226-227	HD 678-679, v. 60, 58-2, (4686).
228-231	HD 680, 724-726, v. 61, 58-2, (4687).
232-236	HD 727-731, v. 62, 58-2, (4688).
237-241	HD 771-775, v. 63, 58-2, (4689).
242-245	HD 776-779, v. 64, 58-2, (4690).
246-254	HD 387-395, v. 84, 58-3, (4863).
255-259	HD 396-400, v. 85, 58-3, (4864).
260-262	HD 401-403, v. 86, 58-3, (4865).
263-268	HD 452-457, v. 87, 58-3, (4866).
269-273	HD 202-206, v. 73, 59-1, (5013).
274	HD 207, v. 74, 59-1, (5014).
275-280	HD 469, 824-828, v. 75, 59-1, (5015).
281-285	HD 829-833, v. 76, 59-1, (5016).
286-293	HD 834-839, 931-932, v. 77, 59-1, (5017).
294-299	HD 933-938, v. 78, 59-1, (5018).
300-301	HD 53-54, v. 56, 59-2, (5159).
302-307	HD 55-60, v. 57, 59-2, (5160).
308-310	HD 61-63, v. 58, 59-2, (5161).
311-315	HD 781-782, 796-798, v. 59, 59-2, (5162).
316	HD 823, v. 60, 59-2, (5163).
317-321	HD 824-828, v. 61, 59-2, (5164).

MINERAL RESOURCES OF THE UNITED STATES (As published by the U. S. Geological Survey. Earlier volumes in this series are listed in the main body of the bibliography.)

1882	HMD 75, v. 40, 48-1, (2252).
1883-1884	HMD 36, v. 9, 49-1, (2414).
1885	HMD 146, v. 6, 49-2, (2493).
1886	HMD 42, v. 2, 50-1, (2566).
1887	HMD 4, v. 2, 50-2, (2655).
1888	HMD 230, v. 16, 51-1, (2775).
1889-1890	HMD 296, v. 42, 52-1, (3000).
1891	HMD 83, v. 7, 52-2, (3116).
1892	HMD 38, v. 4, 53-2, (3232).

1893	HMD 181, v. 28, 53-2, (3256).
1895-1899	(Published as part of the Annual Reports)
1900	HD 17, v. 50, 57-1, (4317).
1901	HD 17, v. 44, 57-2, (4483).
1902	HD 658, v. 92, 58-2, (4718).
1903	HD 20, v. 49, 58-3, (4828).
1904	HD 21, v. 42, 59-1, (4982).
1905	HD 21, v. 40, 59-2, (5143).

MONOGRAPHS OF THE U. S. GEOLOGICAL SURVEY

1	HMD 194, v. 17, 51-1, (2776).
2	HMD 35, v. 11, part 1-2, 48-2, (2320-2321).
3	HMD 52, v. 17, 47-1, (2051-2052).
4	HMD 51, v. 16, 47-1, (2050).
5	HMD 50, v. 15, 47-1, (2049).
6	HMD 43, v. 14, 47-2, (2153).
7	HMD 72, v. 37, 48-1, (2249).
8	HMD 73, v. 38, 48-1, (2250).
9	HMD 74, v. 39, 48-1, (2251).
10-11	HMD 305, 304, v. 2, 49-1, (2407).
12	HMD 397, v. 27, part 1-2, 49-1, (2433-2434).
13	HMD 610, v. 24, part 1-2, 50-1, (2593-2594).
14	HMD 611, v. 25, 50-1, (2595).
15	HMD 147, v. 17, 50-2, (2671).
16	HMD 249, v. 37, 51-1, (2796).
17	HMD 60, v. 29, 52-1, (2987).
18	HMD 77, v. 30, 52-1, (2988).
19	HMD 342, v. 52, 52-1, (3039).
20	HMD 343, v. 53, part 1-2, 52-1, (3040-3041).
21	HMD 118, v. 27, 52-2, (3136).
22	HMD 47, v. 5, 53-2, (3233).
23	HMD 119, v. 14, 53-2, (3242).
24	HMD 120, v. 15, 53-2, (3243).
25	HD 297, v. 63, 54-1, (3430).
26	HD 411, v. 76, 54-1, (3443).
27	HD 136, v. 45, 54-2, (3521).
28	HD 218, part 1-2, v. 53-54, 54-2, (3529-3530).
29	HD 581, v. 82, 55-2, (3711).
30	HD 582, v. 83, 55-2, (3712).
31	HD 101, v. 55-56, 55-3, (3797-3798).
32	HD 287, part 1-3, v. 57-58, part 1-2, 55-3, (3799-3800.2).
33	HD 288, v. 59, 55-3, (3801).
34	HD 289, v. 60, 55-3, (3802).
35	HD 290, v. 61, 55-3, (3803).
36	HD 291, v. 62, 55-3, (3804).
37	HD 292, v. 63, 55-3, (3805).

Publications in the Congressional Documents Set 11

38	HD 315, v. 64, 55-3, (3806).
39	HD 726, v. 108, 56-1, (4005).
40	HD 743, v. 109, 56-1, (4006).
41	HD 310, v. 91, 57-1, (4358).
42	HD 219, v. 73, 57-2, (4512).
43	HD 429, v. 74, 57-2, (4513).
44	HD 432, v. 75, 57-2, (4514).
45	HD 433, v. 76-77, 57-2, (4515-4516).
46	HD 565, v. 82, 58-2, (4708).
47	HD 753, v. 83, 58-2, (4709).
48	HD 475, v. 114-115, 58-3, (4893-4894).
49-50	HD 310-311, v. 68, 59-2, (5171).

PROFESSIONAL PAPERS OF THE U. S. GEOLOGICAL SURVEY

1-6	HD 209-214, v. 68, 57-2, (4507).
7-10	HD 215-218, v. 69, 57-2, (4508).
11-12	HD 448-449, v. 70, 57-2, (4509).
13-14	HD 450-451, v. 71, 57-2, (4510).
15-19	HD 478-482, v. 72, 57-2, (4511).
20-22	HD 566, 709-710, v. 84, 58-2, (4710).
23-26	HD 711-714, v. 85, 58-2, (4711).
27-31	HD 715-719, v. 86, 58-2, (4712).
32-34	HD 780-782, v. 87, 58-2, (4713).
35-37	HD 407-409, v. 90, 58-3, (4869).
38-39	HD 410-411, v. 91, 58-3, (4870).
40-41	HD 196-197, v. 64, 59-1, (5004).
42-43	HD 198-199, v. 65, 59-1, (5005).
44-45	HD 200-201, v. 66, 59-1, (5006).
46-47	HD 488-489, v. 67, 59-1, (5007).
48	HD 490, v. 68-70, 59-1, (5008-5010).
49-53	HD 491-493, 919-920, v. 71, 59-1, (5011).
54-55	HD 921-922, v. 72, 59-1, (5012).
56-57	HD 784-785, v. 69, 59-2, (5172).
58	HD 815, v. 70, 59-2, (5173).

WATER SUPPLY PAPERS OF THE U. S. GEOLOGICAL SURVEY

1-9	HD 108, 342-346, 349-351, v. 70, 54-2, (3546).
10-18	HD 301-302, 453, 477, 509-511, 579-580, v. 60, 55-2, (3688).
19-30	HD 220, 277-283, 297-300, v. 73, 55-3, (3815).
31-39	HD 119, 350-351, 663, 721-725, v. 73, 56-1, (3970).
40-52	HD 106, 484-488, 530-534, 553-554, v. 63, 56-2, (4137).

53-64	HD 53-60, 481-484, v. 61, 57-1, (4328).
65-73	HD 684-692, v. 62, 57-1, (4329).
74-77	HD 200-203, v. 61, 57-2, (4500).
78-81	HD 441-444, v. 62, 57-2, (4501).
82-85	HD 445, 475, 488-489, v. 63, 57-2, (4502).
86-88	HD 490-492, v. 64, 57-2, (4503).
89-92	HD 688-691, v. 93, 58-2, (4719).
93-96	HD 692-695, v. 94, 58-2, (4720).
97-98	HD 735-736, v. 95, 58-2, (4721).
99-100	HD 737-738, v. 96, 58-2, (4722).
101-104	HD 739-742, v. 97, 58-2, (4723).
105-109	HD 759-763, v. 98, 58-2, (4724).
110-116	HD 764-770, v. 99, 58-2, (4725).
117-123	HD 415-420, v. 94, 58-3, (4873).
124-130	HD 432-438, v. 95, 58-3, (4874).
131-135	HD 439-443, v. 96, 58-3, (4875).
136-143	HD 444-451, v. 97, 58-3, (4876).
144-149	HD 461-466, v. 98, 58-3, (4877).
150-156	HD 231-233, 551-554, v. 79, 59-1, (5019).
157-164	HD 555-556, 743-748, v. 80, 59-1, (5020).
165-172	HD 749-756, v. 81, 59-1, (5021).
173-178	HD 757-762, v. 82, 59-1, (5022).
179-183	HD 64-68, v. 62, 59-2, (5165).
184-191	HD 69-76, v. 63, 59-2, (5166).
192-193	HD 786-787, v. 64, 59-2, (5167).
194-195	HD 788-789, v. 65, 59-2, (5168).
196-197	HD 790-791, v. 66, 59-2, (5169).
198-200	HD 792-794, v. 67, 59-2, (5170).
201-206	HD 764-769, v. 38, 60-1, (5307).
207-212	HD 770-775, v. 39, 60-1, (5308).
213-218	HD 776-780, 915, v. 40, 60-1, (5309).
219-223	HD 1019-1022, 1294, v. 64, 60-2, (5474).
224-230	HD 1295-1297, 1524-1526, 1534, v. 65, 60-2, (5475).
231-234	HD 1535, 1555-1557, v. 66, 60-2, (5476).
235-240	HD 166-171, v. 60, 61-2, (5763).
241-244	HD 172-173, 786-787, v. 61, 61-2, (5764).
245-249	HD 788-792, v. 62, 61-2, (5765).
250-252	HD 793-795, v. 63, 61-2, (5766).
253-255	HD 994, 984-985, v. 64, (5767).
256	HD 1034, v. 61, 61-3, (6005).
257-260	HD 1048, 1046, 1449, 1029, v. 62, 61-3, (6006).
261-263	HD 1450, 1021, 1051, v. 63, 61-3, (6007).
264-266	HD 1020, 1390, 1430, v. 64, 61-3, (6008).
267-269	HD 1431, 1426, 1451, v. 65, 61-3, (6009).
270-271	HD 1044, 1455, v. 66, 61-3, (6010).
272	HD 1429, v. 67, 61-3, (6011).

273-274	HD 1360, 1410, v. 68, 61-3, (6012).
275-278	HD 1440, 1443, 1456, 1459, v. 69, 61-3, (6013).
279-280	HD 1489-1490, v. 70, 61-3, (6014).
281-283	HD 1491-1493, v. 71, 61-3, (6015).
284-286	HD 1494-1496, v. 72, 61-3, (6016).
287-289	HD 1497-1498, v. 73, 61-3, (6017).
290-291	HD 1511-1512, v. 74, 61-3, (6018).
292	HD 1513, v. 75, 61-3, (6019).
293	HD 666, v. 62, 62-2, (6244).
294	HD 737, v. 63, 62-2, (6245).
295-297	HD 738, 813, 836, v. 64, 62-2, (6246).
298	HD 837, v. 65, 62-2, (6247).
299	HD 838, v. 66, 62-2, (6248).
300	HD 839, v. 67, 62-2, (6249).
301-303	HD 840-842, v. 68, 62-2, (6250).
304-306	HD 843, 883-884, v. 69, 62-2, (6251).
307-309	HD 885-887, v. 70, 62-2, (6252).
310-311	HD 888-889, v. 71, 62-2, (6253).
312	HD 547, v. 72, 62-2, (6254).
313	HD 980, v. 55, 62-3, (6421).
314	HD 981, v. 56, 62-3, (6422).
315-317	HD 1236-1238, v. 57, 62-3, (6423).
318	HD 1424, v. 58, 62-3, (6424).
319	HD 1425, v. 59, 62-3, (6425).
320	HD 1454, v. 60, 62-3, (6426).

Part 3

BIBLIOGRAPHY

Abercrombie, W. R., 1900, Copper River exploring expedition, SD 306, v. 32, 56-1, p. 1-169, 168 plates, 1 map, (3874).

Describes the geology and economic deposits of this part of Alaska.

Abert, J. W., 1846, Report of an expedition on the Upper Arkansas and through the country of the Camanche Indians in the fall of the year 1845, SED 438, v. 8, 29-1, p. 2-75, 11 plates, 2 maps, (477).

Describes the landforms and some of the lithology of the area explored.

Abert, J. W., 1848, . . . Examination of New Mexico in the years 1846 to 1847, SED 23, v. 4, 30-1, p. 3-132, 24 plates, 1 map, (506).

Also in HED 41, v. 4, 30-1, p. 417-548, 24 plates, 1 map, (517). Describes the landforms and lithology of the area. Three plates illustrate fossils. See also Emory, William H., 1848.

Abbot, Henry L., 1855, Report upon exploration for a railroad route from the Sacramento Valley to the Columbia River, SED 78, v. 13, part 6, 33-2, p. 1-134, 13 plates, (763).

Also in HED 91, v. 11, part 6, 33-2, p. 1-134, 13 plates, (796). Part of the Pacific Railroad Survey Reports. Describes the landforms and lithology, see also Newberry, J. S., 1855.

Adam, J. S., 1871, Notes on the geology of the vicinity of Samana, SED 9, v. 1, 42-1, p. 70-71, (1466).

Includes a brief description of the areal geology and is one of a number of reports describing the Dominican Republic.

Adams, Charles C., 1905, The post-glacial dispersal of the North American biota, HD 460, v. 111, 58-3, p. 623-637, (4890).

In the Report of the 8th International Geographic Congress.

Adams, Samuel, 1870, Letters and report on exploration of the Colorado River, HMD 12, v. 1, 41-3, p. 1-20, (1462).

Also in HMD 37, v. 1, 42-1, p. 1-20, (1472). Includes a journal of a supposed trip down the river and describes the landforms and mineral resources of the area. Adams has been described as "one of a tribe of Western adventurers and imposters and mountebanks" (W. Stegner, *Beyond the Hundredth Meridian*).

Agassiz, Alexander, 1881, Dredging operations in the Caribbean Sea, SED 17, v. 2, 46-2, p. 95-102, (1883).

Describes the sediments and bottom features of parts of the Caribbean. In the Annual Report of the Coast and Geodetic Survey.

Agassiz, Louis, 1852, Extracts from the report . . . on the examination of the Florida reefs, keys, and coast, SED 3, v. 5, 32-1, p. 145-160, (615).

Also in HED 87, v. 14, 39-2, p. 120-130, (1296). Briefly describes the coastal features, reefs, and sediments.

Agassiz, Louis, 1855, Notice of the fossil fishes (found while exploring routes in California to connect with routes near the 35th and 32nd parallels), SED 78, v. 13, part 5, 33-2, p. 313-316, 1 plate, (762).

Also in HED 91, v. 11, part 5, 33-2, p. 313-316, 1 plate, (795). Part of the Pacific Railroad Survey Reports. In Williamson, R.S., 1855.

Agassiz, Louis, 1868, Letters on the relation of geological and zoological researches to general interests, in the development of coast features, HED 275, v. 18, 40-2, p. 183-186, (1344).

Emphasizes the importance of the study of the degrading and aggrading portions of the U. S. coastline. In the Annual Report of the Coast Survey.

Agassiz, Louis, 1872a, (Instructions for the study of glaciers on the expedition to the North Pole), SMD 149, v. 2, 42-2, p. 385-387, (1482).

Also in HED 1, part 3, v. 4, 42-2, p. 258-260, (1507). In the Annual Report of the Smithsonian Institution.

Agassiz, Louis, 1872b, Report upon deep sea dredgings in the Gulf Stream . . . , HED 206, v. 8, 41-2, p. 208-219, (1419).

Describes the bottom sediments found east of Florida. In the Annual Report of the Coast Survey.

Alabama, University of, 1902, Memorial of the University of Alabama for opening the navigable rivers that drain the coal and iron fields of Alabama, SD 161, v. 13, 57-1, p. 1-13, (4231).

Describes the coal and iron deposits of parts of Alabama.

Allen, Henry T., 1887, Report of an expedition to the Copper, Tanana, and Koyokuk Rivers in the Territory of Alaska in the year 1885, SED 125, v. 2, 49-2, p. 1-172, 29 plates, 5 maps, (2449).

Describes the exploration, landforms, and general geology of the areas.

Allen, J., 1834, Letter and journal of an expedition into the indian country to the source of the Mississippi River, HED 323, v. 4, 23-1, p. 1-68, 1 map, (257).

Contains detailed landform descriptions and generalized rock descriptions. Ancient strandlines of Lake Superior and copper deposits are also described.

Allen, J., 1846, Journal of a march into the indian country in the northern part of Iowa Territory in 1844, by Company I, 1st Regiment of Dragoons, HED 168, v. 6, 29-1, p. 7-18, (485).

Describes the exploration and landforms and briefly mentions the rock types found.

Anderson, Martin B., 1871, Sketch of the life of Prof. Chester Dewey, HED 20, v. 2, 42-1, p. 231-240, (1471).

A review of the life and work of this American mineralogist. In the Annual Report of the Smithsonian Institution.

Anderson, Richard J., 1905, The flora of Connaught as evidence of the former connection with an Atlantic continent, HD 460, v. 111, 58-3, p. 613-615, (4890).

Describes the Quaternary plant distribution of parts of Ireland and North America. In the Report of the 8th International Geographic Congress.

Anderson, Tempest, and **Flett, John S.**, 1903, Preliminary report on the recent eruption of the Soufrière in St. Vincent, and of a visit to Mont Pelée, in Martinique, HD 484, part 1, v. 109, 57-2, p. 309-330, 3 plates, (4548).

Describes the results of the famous eruption of Pelée. In the Annual Report of the Smithsonian Institution. See also Russell, Israel, C., 1903.

Andrews, C. W., 1907, The recently discovered Tertiary vertebrata of Egypt, HD 575, v. 97, 59-2, p. 295-307, (5200).

In the Annual Report of the Smithsonian Institution.

Anthony, Raoul, 1904, The evolution of the human foot, HD 748, v. 114, 58-2, p. 519-535, (4740).

In the Annual Report of the Smithsonian Institution.

Antisell, Thomas, 1855, Geological report on routes in California to connect with routes near the 35th and 32nd parallels, and the route . . . between the Rio Grande and the Pimas villages, SED 78, v. 13, part 7, 33-2, p. 1-204, 24 plates, 2 maps, (764).

Also in HED 91, v. 11, part 7, 33-2, p. 1-204, 24 plates, 2 maps, (797). This describes the geology of the routes surveyed and is part of the Pacific Railroad Survey Reports. In Parke, John G., 1855a.

Arago, D. F. J., 1872, Biography of Joseph Fourier, SMD 149, v. 2, 42-2, p. 137-176, (1482).

A review of the life and work, including paleontology, of Fourier. In the Annual Report of the Smithsonian Institution.

Arctowski, Henryk, 1902, The antarctic voyage of the Belgica during the years 1897, 1898, and 1899, HD 707, part 1, v. 126, 57-1, p. 377-388, 6 plates, 1 map, (4393).

Describes the glacial features and landforms of parts of the continent. In the Annual Report of the Smithsonian Institution.

Ashburner, William, 1867, Geological formation etc. of the Pacific slope, HED 29, v. 7, 39-2, p. 37-49, (1289).

Describes the geology and mineral deposits of the California gold area. In Browne, J. R., 1867.

Ashburner, William, 1869, Report on certain mines in Idaho, HED 54, v. 9, 40-3, p. 161-164, (1374).

In Raymond, Rossiter W., 1869.

Avon-Burke, R. T., 1904a, Soil survey of Mason County, Kentucky, HD 746, v. 111, 58-2, p. 631-645, 1 map (in v. 112), (4737-4738).

Avon-Burke, R. T., 1904b, Soil survey of Scott County, Kentucky, HD 746, v. 111, 58-2, p. 619-630, 1 map (in v. 112), (4737-4738).

Avon-Burke, R. T., and **Marean, Herbert,** 1902a, Soils survey of Cobb County, Georgia, HD 655, v. 117, 57-1, p. 317-327, 1 map, (4384).

Avon-Burke, R. T., and **Marean, Herbert,** 1902b, Soils survey of the Westfield area, New York, HD 655, v. 117, 57-1, p. 75-92, 5 plates, 1 map, (4384).

Avon-Burke, R. T., and **Root, Aldert S.,** 1907, Soil survey of Yorktown area, Virginia, HD 925, v. 110, 59-1, p. 247-270, 1 map (in v. 111), (5050-5051).

Avon-Burke, R. T., and **Ruhlen, La Mott,** 1904a, Soil survey of Madison County, Indiana, HD 746, v. 111, 58-2, p. 687-701, 1 map (in v. 112), (4737-4738).

Avon-Burke, R. T., and **Ruhlen, La Mott,** 1904b, Soil survey of Shelby County, Missouri, HD 746, v. 111, 58-2, p. 875-889, 1 map (in v. 112), (4737-4738).

Avon-Burke, R. T., and **Wilder, Henry J.,** 1903, Soil survey of the Trenton area, New Jersey, HD 473, v. 106, 57-2, p. 163-186, 1 map, (4545).

Avon-Burke, R. T., et al, 1903, Soil survey of Perry County, Alabama, HD 473, v. 106, 57-2, p. 309-323, 1 map, (4545).

Avon-Burke, R. T., et al, 1904, Soil survey of the Mobile area, Alabama, HD 746, v. 111, 58-2, p. 393-403, 1 map (in v. 112), (4737-4738).

Babinet, Jacques, 1871, The diamond and other precious stones, HED 20, v. 2, 42-1, p. 333-363, (1471).

Describes gems and the history of gems. In the Annual Report of the Smithsonian Institution.

Bache, A. D., 1856a, Notice of earthquake waves on the western coast of the United States, on the 23rd and the 25th December, 1854, SED 22, v. 17, 34-1, p. 342-346, (826).

Also in HED 6, v. 3, 34-1, p. 342-346, (845), *and* HED 22, v. 9, 37-3, p. 238-241, (1165). Describes the effects of a Pacific Coast tsunami. In the Annual Report of the Coast Survey.

Bache, A. D., 1856b, Observations to determine the cause of the increase of Sandy Hook, . . . New York, SED 18, v. 15, 34-3, p. 263-264, (888).

Also in HED 18, v. 4, 34-3, p. 263-264, (898). Describes shore modification by sand deposition. In the Annual Report of the Coast Survey.

Bailey, J. W., 1845, (Report on fresh water infusoria from Oregon), SED 174, v. 11, 28-2, p. 302, 1 plate, (461).

Also in HED 166, v. 4, part 2, 28-2, p. 302, 1 plate, (466). Describes some fossil diatoms collected by the Fremont expedition. In Hall, James, 1845b.

Bailey, J. W., 1848, Notes concerning the fossils collected by Lieutenant J. W. Abert, while engaged in the geographical examination of New Mexico, SED 23, v. 4, 30-1, p. 131-132, 3 plates, (506).

Also in HED 41, v. 4, 30-1, p. 547-548, 3 plates, (517). Contains short descriptions of the rocks and fossils collected. In Abert, J. W., 1848.

Bailey, J. W., 1855a, Letter describing the structure of the fossil plant from Posuncula River, SED 78, v. 13, part 5, 33-2, p. 337, 1 plate, (762).

Also in HED 91, v. 11, part 5, 33-2, p. 337, 1 plate, (795). Part of the Pacific Railroad Survey Reports. In Williamson, R. S., 1855.

Bailey, J. W., 1855b, Letter upon infusorial fossils submitted . . . by Dr. Schiel, SED 78, v. 13, part 2, 33-2, p. 111-112, (759).

Also in HED 91, v. 11, part 2, 33-2, p. 111-112, (792), *and* HED 129, v. 18, part 2, 33-1, p. 135-136, (737). Describes some diatoms collected in California. Part of the Pacific Railroad Survey Reports. In Beckwith, E. G., 1855a.

Bailey, J. W., 1856a, . . . characteristics deducible from specimens of bottom, brought up in sounding the Florida section of the Gulf Stream, SED 22, v. 17, 34-1, p. 360, (826).

Also in HED 6, v. 3, 34-1, p. 360, (845). Describes deep water bottom samples. The included shallow water organic remains are interpreted as deposited, at least in part, by currents. In the Annual Report of the Coast Survey.

Bailey, J. W., 1856b, (Description of bottom sediments obtained by the "Arctic" while surveying a route for a telegraphic cable from Newfoundland to Ireland), SED 5, v. 3, 34-3, p. 469-472, (876).

Also in HED 1, v. 1, part 2, 34-3, p. 469-472, (894). Contains brief descriptions of some deep water sediments. In the Annual Report of the Secretary of the Navy.

Bain, H. Foster, 1904, Reported gold deposits of the Wichita Mountains, SD 149, v. 4, 58-2, p. 1-10, (4589).

Describes the geology and the gold mines of the Wichita Mts., Oklahoma.

Baird, Spencer F., 1852, An account of natural history explorations in the United States during 1851, SMD 108, v. 1, 32-1, p. 52-56, (629).

Contains brief accounts of the various explorations then underway. In the Annual Report of the Smithsonian Institution.

Baird, Spencer F., 1853, Account of scientific explorations, and reports on explorations, made in America during the year 1852, SMD 53, v. 1, 32-2, p. 58-65, (670).

Briefly describes the explorations then underway. In the Annual Report of the Smithsonian Institution.

Baird, Spencer F., 1855, Report on American explorations in the years 1853 and 1854, SMD 24, v. 1, 33-2, p. 79-97, (772).

Also in HMD 37, v. 1, 33-2, p. 79-97, (807). Briefly describes the explorations then underway. In the Annual Report of the Smithsonian Institution.

Bannister, H. M., 1873, Report of a geological reconnoissance along the Union Pacific Railroad, HMD 112, v. 3, 42-3, p. 521-541, (1573).

Describes the geology of parts of Wyoming. In Hayden, F. V., 1873.

Bannon, M., and **Walcott, Charles D.,** 1884, Deer Creek coal field, White Mountain Indian Reservation, Arizona, SED 20, v. 1, 48-2, p. 2-7, 1 plate, (2261).

Describes the geology of the area and includes a cross-section.

Barlow, J. W., 1872, Report of a reconnaissance of the Yellowstone River in 1871, SED 66, v. 2, 42-2, p. 1-42, (1479).

In Wheeler, George M., 1872a.

Barnard, E. C., 1899, Report of the Forty Mile Expedition, HD 172, v. 13, 55-3, p. 76-84, (3737).

Describes the geology and economic deposits of parts of Alaska.

Barnes, George Orville, 1849a, Field notes (on the Ontonagon district), SED 1, v. 3, 31-1, p. 627-636, (551).

Also in HED 5, v. 3, part 3, 31-1, p. 627-636, (571). In Jackson, Charles T., 1849.

Barnes, George Orville, 1849b, Report (on the Lake Superior region), SED 1, v. 3, 31-1, p. 509-514, (551).

Also in HED 5, v. 3, part 3, 31-1, p. 509-514, (571). In Jackson, Charles T., 1849.

Bauer, L. A., 1905, The magnetic disturbances during the eruption of Mount Pelée on May 8, 1902, HD 460, v. 111, 58-3, p. 501-502, (4890).

In the Report of the 8th International Geographic Congress.

Beale, Edward F., 1858, Wagon road from Fort Defiance to the Colorado River, HED 124, v. 13, 35-1, p. 1-87, 1 map, (959).

Describes an exploration from Texas to the Colorado River. Contains landform descriptions and brief rock descriptions.

Beale, Edward F., and **Engle, F. E.,** 1860, Report . . . relating to the construction of a wagon road from Fort Smith to the Colorado, HED 42, v. 6, 36-1, p. 1-91, 1 map, (1048).

Includes topographic description and brief rock descriptions.

Becker, George F., 1901, Mineral resources and geology (of the Philippines), SD 138, v. 45, 56-1, p. 229-240, (3885, v. 2).

Describes the general geology and the known resources.

Beckwith, E. G., 1855a, Report of exploration for the Pacific Railroad, on the line of the 41st parallel of north latitude, SED 78, v. 13, part 2, 33-2, p. 1-132, 14 plates, (759).

Also in HED 91, v. 11, part 2, 33-2, p. 1-132, 14 plates, (792). Also in HED 129, v. 18, part 2, 33-1, p. 1-136, (737). Part of the Pacific Railroad Survey Reports. Describes the landforms and the lithology.

Beckwith, E. G., 1855b, Report of exploration of a route for the Pacific Railroad near the 38th and 39th parallels of latitude from the mouth of the Kansas to the Sevier River, in the Great Basin, SED 78, v. 13, part 2, 33-2, p. 1-128, 13 plates, (759).

Also in HED 91, v. 11, part 2, 33-2, p. 1-128, 13 plates. (792). Also in HED 129, v. 18, part 2, 33-1, p. 1-149, (737). Part of the Pacific Railroad Survey Reports. Describes parts of one of the possible central routes and contains a great deal of rock and landform description.

Bell, Robert, 1898, Rising of the land around Hudson Bay, HD 575, part 1, v. 78, 55-2, p. 359-367, (3706).

Describes the glacial features and isostatic rebound of parts of Ontario and Quebec. In the Annual Report of the Smithsonian Institution.

Bell, William H., 1844, Report on the mineral lands of the upper Mississippi, HED 43, v. 3, 28-1, p. 1-52, (441).

Describes in detail the metaliferous deposits and compares these with similar deposits in England.

Bennett, Frank, Jr., 1904, Soil survey of the Brookings area, South Dakota, HD 746, v. 111, 58-2, p. 963-977, 1 map (in v. 112), (4737-4738).

Bennett, Frank, Jr., and **Ely, Charles W.,** 1905, Soil survey of Marshall County, Indiana, HD 458, v. 108, 58-3, p. 689-706, 1 map (in v. 109), (4887-4888).

Bennett, Frank, Jr., and **Griffen, A. M.,** 1904, Soil survey of the Huntsville area, Alabama, HD 746, v. 111, 58-2, p. 373-392, 1 map (in v. 112), (4737-4738).

Bennett, Frank, Jr., and **Griffen, A. M.,** 1905, Soil Survey of the Orangeburg area, South Carolina, HD 458, v. 108, 58-3, p. 185-205, 1 map (in v. 109), (4887-4888).

Bennett, Frank, Jr., and **Jones, Grove B.,** 1903, Soil survey of the Brazoria area, Texas, HD 473, v. 106, 57-2, p. 349-364, 2 plates, 1 map, (4545).

Bennett, Frank, Jr., and **Winston, R. A.,** 1908, Soil survey of Pontotoc County, Mississippi, HD 352, v. 75, 59-2, p. 405-426, 1 map (in v. 76), (5178-5179).

Bennett, Frank, Jr., et al, 1908, Soil survey of Chesterfield County, Virginia, HD 352, v. 75, 59-2, p. 195-222, 1 map (in v. 76), (5178-5179).

Bennett, Hugh H., and **Hurst, Lewis A.,** 1908, Soil survey of Blue Earth County, Minnesota, HD 352, v. 75, 59-2, p. 813-863, 1 map (in v. 76), (5178-5179).

Bennett, Hugh H., and **McLendon, W. E.,** 1907a, Soil survey of Hanover County, Virginia, HD 925, v. 110, 59-1, p. 213-245, 1 map (in v. 111), (5050-5051).

Bennett, Hugh H., and **McLendon, W. E.,** 1907b, Soil survey of Louisa County, Virginia, HD 925, v. 110, 59-1, p. 191-212, 1 map (in v. 111), (5050-5051).

Bernard, S., and Totten, Joseph G., 1823, Report of the Board of Engineers on the Ohio and Mississippi Rivers made in the year 1821, HED 35, v. 3, 17-2, p. 7-22, (78).

Describes the channel geometry of the two rivers and includes a general discussion of floods. The plans, bar surveys, and sketches of the original report are not included here.

Bixby, William H., Beach, Lansing, H., and **Gaillard, D. D.,** 1906, Effect of wave action at certain harbors on Lake Michigan, HD 62, v. 44, 59-1, p. 1-24, 1 map, (4984).

Describes the lacustrine features and shoreline changes of parts of Michigan and Wisconsin.

Blake, Theodore A., 1869, General topographical and geological features of the northwestern coast of America, from the straits of Juan de Fuca to the parallel of 60 degrees north latitude, HED 275, v. 18, 40-2, p. 281-290, (1344).

In the Annual Report of the Coast Survey.

Blake, William P., 1855a, General report on the geology of the route (near the 35th parallel), SED 78, v. 13, part 3, 33-2, p. 1-175, 5 plates, 2 maps, (760).

Also in HED 91, v. 11, part 3, 33-2, p. 1-175, 5 plates, 2 maps, (793). Part of the Pacific Railroad Survey Reports. Found in Whipple, A. W., 1855.

Blake, William P., 1855b, Geological report on the routes in California to connect with the routes near the 35th and 32nd parallels, SED 78, v. 13, part 5, 33-2, p. 1-310, 20 plates, 4 maps, (762).

Also in HED 91, v. 11, part 5, 33-2, p. 1-310, 20 plates, 4 maps, (795). Part of the Pacific Railroad Survey Reports. In Williamson, R. S., 1855.

Blake, William P., 1855c, Report on the geology of the route, near the 32nd parallel, SED 78, v. 13, part 2, 33-2, p. 1-50, 1 plate, 1 map, (759).

Also in HED 91, v. 11, part 2, 33-2, p. 1-50, 1 plate, 1 map, (792). Part of the Pacific Railroad Survey Reports. This describes the geology of the proposed southern route. It was supposed to be published in Pope, John, 1855, but was instead published separately.

Blake, William P., 1856a, Notice of the geological collection (of an expedition to the sources of the Brazos and Big Witchita Rivers), SED 60, v. 12, 34-1, p. 46-47, (821).

Briefly describes specimens collected during an exploration of this part of Texas. In Marcy, Randolph B., 1856.

Blake, William P., 1856b, Observations on the physical geography of the coast of California, from Bodega Bay to San Diego, SED 22, v. 17, 34-1, p. 376-398, 4 maps, (826).

Also in HED 6, v. 3, 34-1, p. 376-398, 4 maps, (845). Briefly describes the topography of the California coast. In the Annual Report of the Coast Survey.

Blake, William P., 1867, Annotated catalogue of the principal mineral species hitherto recognized in California, HED 29, v. 7, 39-2, p. 200-211, (1289).

In Browne, J. R., 1867.

Blake, William P., 1869, Report upon the precious metals . . . of the principal gold and silver producing regions of the world represented at the Paris Universal Exposition, HMD—un-numbered, v. 2, 40-2, p. 1-369, (1352).

In the Reports of the U. S. Commissioners to the Paris Universal Exposition. Describes in detail the history, geology, and mining of gold, silver, and platinum.

Blake, William P., 1871a, Preliminary report of observations upon the peninsula of Samana, SED 9, v. 1, 42-1, p. 63-70, (1466).

Briefly describes the geology of this portion of the Dominican Republic.

Blake, William P., 1871b, Preliminary report of the expedition across the island from Santo Domingo City to Puerto Plata . . . , SED 9, v. 1, 42-1, p. 121-127, (1466).

Briefly describes the geology of this part of the Dominican Republic.

Blake, William P., 1871c, Preliminary report upon the mineral resources of Santo Domingo, SED 9, v. 1, 42-1, p. 144-145, (1466).

Briefly describes the known resources of the Dominican Republic.

Blake, William P., 1885, The various forms in which gold occurs in nature, HED 268, v. 33, 48-2, p. 573-597, (2308).

Describes the occurrence, crystalography, and other properties of gold.

Blake, William P., 1901, Geology of Arizona, HD 5, v. 26, 57-1, p. 194-198, (4293).

A brief description of the general geology of the State.

Blake, William P., 1903, Geology of Arizona, HD 5, v. 22, 58-2, p. 226-235, (4648).

Describes the general geology and in particular the stratigraphy of the state.

Blanchard, I., 1869, Mineral resources of the Isthmus of Panama, HED 54, v. 9, 40-3, p. 169-172, (1374).

In Raymond, Rossiter W., 1869.

Blatchly, A., 1867, Report on mining in southeastern Nevada, HED 29, v. 7, 39-2, p. 132-135, (1289).

Describes the mining and areal geology of southeastern Nevada. In Browne, J. R., 1867.

Blytt, A., 1890, On movements of the earth's crust, HMD 224, part 1, v. 20, 51-1, p. 325-375, (2779).

Describes cycles of tectonic and epirogenic movements and discusses the relation of these to past climates. In the Annual Report of the Smithsonian Institution.

Boehmer, George H., 1886, Volcanic eruptions and earthquakes in Iceland within historic times, HMD 15, v. 25, 49-1, p. 495-541, (2431).

A catalog and description of the known eruptions and earthquakes. In the Annual Report of the Smithsonian Institution.

Bonsteel, F. E., and Ayrs, O. L., 1904, Soil survey of the Dover area, Delaware, HD 746, v. 111, 58-2, p. 143-164, 1 map (in v. 112), (4737-4738).

Bonsteel, F. E., and Carr, E. P., 1905a, Soil survey of the Charleston area, South Carolina, HD 458, v. 108, 58-3, p. 207-230, 1 map (in v. 109), (4887-4888).

Bonsteel, F. E., and Carr, E. P., 1905b, Soil Survey of Rhode Island, HD 458, v. 108, 58-3, p. 47-72, 2 maps (in v. 109), (4887-4888).

Bonsteel, F. E., and Carter, William T., Jr., 1904, Soil survey of Worchester County, Maryland, HD 746, v. 111, 58-2, p. 165-189, 1 map (in v. 112), (4737-4738).

Bonsteel, F. E., Carter, William T., Jr., and Ayrs, O. L., 1904, Soil survey of the Syracuse area, New York, HD 746, v. 111, 58-2, p. 63-89, 1 map (in v. 112), (4737-4738).

Bonsteel, F. E., et al, 1907, Soil survey of Lauderdale County, Alabama, HD 925, v. 110, 59-1, p. 389-405, 1 map (in v. 111), (5050-5051).

Bonsteel, Jay A., 1901a, Soil survey of Kent County, Maryland, HD 526, v. 107, 56-2, p. 173-186, 1 map, (4181).

Bonsteel, Jay A., 1901b, Soil survey of St. Mary County, Maryland, HD 526, v. 107, 56-2, p. 125-145, 1 map, (4181).

Bonsteel, Jay A., 1903, Soil survey of the Janesville area, Wisconsin, HD 473, v. 106, 57-2, p. 549-570, 3 plates, 1 map, (4545).

Bonsteel, Jay A., and **Avon-Burke, R. T.,** 1901, Soil survey of Calvert County, Maryland, HD 526, v. 107, 56-2, p. 147-171, (4181).

Bonsteel, Jay A., Fippin, Elmer O., and **Carter, William T.,** 1907, Soil survey of Tomkins County, New York, HD 925, v. 110, 59-1, p. 39-70, 1 map (in v. 111), (5050-5051).

Bonsteel, Jay A., and **Taylor, F. W.,** 1902, Soil survey of the Salem area, New Jersey, HD 655, v. 117, 57-1, p. 125-148, 1 map, (4384).

Bonsteel, Jay A., et al, 1902a, Soil survey of Prince George County, Maryland, HD 655, v. 117, 57-1, p. 173-210, 5 plates, 1 map, (4384).

Bonsteel, Jay A., et al, 1902b, Soil survey of the Yazoo area, Mississippi, HD 655, v. 117, 57-1, p. 359-388, 9 plates, 2 maps, (4384).

Bonsteel, Jay A., et al, 1903a, Soil survey of Clinton County, Illinois, HD 473, v. 106, 57-2, p. 491-505, 1 map, (4545).

Bonsteel, Jay A., et al, 1903b, Soil survey of Tazewell County, Illinois, HD 473, v. 106, 57-2, p. 465-489, 1 map, (4545).

Bonsteel, Jay A., et al, 1904, Soil survey of the Long Island area, New York, HD 746, v. 111, 58-2, p. 91-128, 2 maps (in v. 112), (4737-4738).

Boutelle, Charles O., 1887, Report concerning the earliest topographic surveys of Monomy, HED 40, v. 22, 49-2, p. 260-261, 1 map, (2481).

Describes the shore features of parts of Massachusetts. In the Annual Report of the Coast and Geodetic Survey.

Bowditch, E. W., 1874, The geological formation of the Sassardi and Morti and San Blas Routes, HMD 113, v. 5, 42-3, p. 151-154, 2 plates, (1575).

Describes parts of Panama. In Selfridge, Thomas O., 1874.

Bowers, Stephen, 1878, Santa Rosa Island, SMD 35, v. 1, 45-2, p. 316-320, (1785).

Describes the geology of this California island.

Bowman, Amos, 1873, The Pliocene Rivers of California, HED 210, v. 9, 42-3, p. 377-389, 5 plates, (1567).

In Raymond, Rossiter W., 1873.

Bowman, Amos, 1875, Geology of the Sierra Nevada in its relation to vein-mining, HED 177, v. 18, 43-1, p. 441-469, (1651).

In Raymond, Rossiter W., 1875.

Bradley, F. H., 1873, (Geological report on Utah, Wyoming, and Idaho), HMD 112, v. 3, 42-3, p. 190-274, 2 maps, (1573).

In Hayden, F. V., 1873.

Brenndecke, F., 1871, On meteorites, HED 153, v. 12, 41-3, p. 417-419, (1460).

Describes meteorites, especially those found in Wisconsin. In the Annual Report of the Smithsonian Institution.

Brewer, William H., et al, 1905, Scientic surveys of the Philippine Islands, SD 178, v. 4, 58-3, p. 21-33, (4766).

Also in SD 145, v. 3, 58-3, p. 2-13, (4765). Describes the geological survey of the Philippines.

Brezina, Aristides, 1873, Explanation of the principles of crystalography and crystalophysics, HMD 107, v. 3, 42-3, p. 233-266, (1573).

In the Annual Report of the Smithsonian Institution.

Briggs, Lyman J., 1900a, Salts as influencing the rate of evaporation of water from soils, HD 399, v. 88, 56-1, p. 184-198, 1 plate, (3985).

Describes experimental work done under the direction of the Soil Survey.

Briggs, Lyman J., 1900b, Some necessary modifications in the method of mechanical analysis as applied to alkali soils, HD 399, v. 88, 56-1, p. 173-183, 1 plate, (3985).

Describes methods of soil analysis as used by the Soil Survey.

Briggs, Lyman J., 1901, Investigations on the physical properties of soils, HD 526, v. 107, 56-2, p. 413-421, (4181).

Describes the physical properties of soils and the means of measuring these properties in various soil types.

Brodie, Alexander O., 1904, Geology of Arizona, HD 5, v. 22, 58-3, p. 60-66, (4801).

Describes the general geology and especially the stratigraphy.

Brooks, Alfred H., 1899, The Yukon District, HD 172, v. 13, 55-3, p. 85-100, (3737).

Describes the geology and the mineral deposits of this part of Alaska.

Brooks, Alfred H., 1905, The geography of Alaska, with an outline of the geomorphology, HD 460, v. 111, p. 204-230, 1 map, (4890).

In the Report of the 8th International Geographic Congress.

Brooks, Alfred H., Richardson, George B., and **Collier, Arthur J.,** 1901, Reconnaissance of Cape Nome and adjacent gold fields of Seward Peninsula, Alaska, HD 547, v. 124, 56-2, p. 1-180, 11 plates, 6 maps, (4198).

Describes the geology and the gold deposits of the area.

Brooks, William K., 1896, The origin of the oldest fossils and the discovery of the bottom of the ocean, HMD 90, part 1, v. 15, 53-3, p. 359-376, (3341).

Discusses the origin and evolution of the Cambrian fauna. In the Annual Report of the Smithsonian Institution.

Brooks, William K., 1901, The lesson of the life of Huxley, HD 537, part 1, v. 114, 56-2, p. 701-711, (4188).

Describes the life and work of T. H. Huxley. In the Annual Report of the Smithsonian Institution.

Brown, C. Newton, 1900, Report upon the mineral wealth of the Big Sandy Valley from Louisa to the head of navigation, HD 2, v. 19, 56-2, p. 3413-3461, (4093).

Describes the geology and coal resources of parts of Kentucky, Virginia, and West Virginia.

Brown, Edwin C., 1905, Report on the Minarets, SD 34, v. 2, 58-3, p. 48-49, (4764).

Describes the geology and mineral resources of this part of California. See also Preston, E. B., 1905.

Browne, J. R., 1867, Letter and report upon the mineral resources of the states and territories west of the Rocky Mountains, HED 29, v. 7, 39-2, p. 7-321, (1289).

Describes in detail the mining and economic deposits of the western part of the country. This report along with Taylor, James W., 1867, forms the first in a series which eventually is taken over by the U. S. Geological Survey and continued as *Mineral Resources of the United States*.

Browne, J. R., 1868, Letter and report upon the mineral resources of the states and territories west of the Rocky Mountains, HED 202, v. 16, 40-2, p. 1-674, (1342).

See previous entry above.

Brush, G. J., 1869, Catalogue of meteorites in the mineralogical collection of Yale College, HED 83, v. 12, 40-3, p. 342-344, (1380).

In the Annual Report of the Smithsonian Institution.

Bryant, Charles, 1870, (Report on Alaska), SED 32, v. 1, 41-2, p. 1-25, (1405).

Describes the topography and mineral resources of the coastal regions.

Bulkley, Charles S., 1868, (Letter describing the Alaskan coast and islands), HMD 131, v. 2, 40-2, p. 1-11, (1350).

Describes the topography and known mineral resources.

Bureau of Immigration, 1901, Mines and minerals, HD 5, v. 27, 57-1, p. 393-411, (4294).

Describes the mineral resources of New Mexico.

Bureau of the American Republics, 1902, United States of Brazilia: a geographical sketch with special reference to economic conditions and prospects of future development, HD 557, v. 90, 57-1, p. 1-233, (4357).

Describes the landforms and mineral deposits of Brazil.

Bureau of the American Republics, 1903, Argentine Republic: a geographical sketch . . . , HD 111, part 2, v. 65, 58-2, p. 1-376, 26 plates, 3 maps, (4691).

Describes the landforms and mineral deposits of Argentina.

Bureau of the American Republics, 1904a, Bolivia: geographical sketch, natural resources . . . , HD 145, part 8, v. 67, 58-3, p. 1-214, 19 plates, (4846).

Describes the landforms and mineral resources of Bolivia.

Bureau of the American Republics, 1904b, Mexico: geographical sketch, natural resources . . . , HD 145, part 5, v. 66, 58-3, p. 1-454, 24 plates, (4845).

Describes the landforms and mineral deposits of Mexico.

Bureau of the American Republics, 1905a, Colombia, salt mines of the Republic, HD 267, part 1, v. 68, 58-3, p. 100-102, (4847).

Describes the salt mines and deposits.

Bureau of the American Republics, 1905b, Colored stones, HD 267, part 11, v. 71, 58-3, p. 1312-1317, (4850).

Describes the gems and gem deposits of Brazil.

Burgess, James L., and Coffey, George N., 1905, Soil survey of the Garden City area, Kansas, HD 458, v. 108, 58-3, p. 895-923, 1 map (in v. 109), (4887-4888).

Burgess, James L., and **Lyman, W. S.,** 1907, Soil survey of Lee County, Texas, HD 925, v. 110, 59-1, p. 601-621, 1 map (in v. 111), (5050-5051).

Burgess, James L., and **Tharp, W. E.,** 1907, Soil survey of the Crystalsprings area, Mississippi, HD 925, v. 110, 59-1, p. 473-491, 1 map (in v. 111), (5050-5051).

Burgess, James L., Tharp, W. E., and **Lyman, W. S.,** 1907, Soil survey of Brown County, Kansas, HD 925, v. 110, 59-1, p. 911-926, 1 map (in v. 111), (5050-5051).

Burgess, James L., and **Worthen, H. L.,** 1908, Soil survey of Lancaster County, Nebraska, HD 352, v. 75, 59-2, p. 943-962, 1 map (in v. 76), (5178-5179).

Burgess, James L., et al, 1908, Soil survey of Caddo Parish, Louisiana, HD 352, v. 75, 59-2, p. 427-458, 1 map (in v. 76), (5178-5179).

Burr, William H., 1903, The Panama route for a ship canal, HD 484, part 1, v. 109, 57-2, p. 537-557, 2 plates, (4548).

Describes topography and geological engineering problems. In the Annual Report of the Smithsonian Institution.

Burt, William A., 1846, Topography and geology of the survey of a district of township lines, south of Lake Superior, 1845, SED 357, v. 7, 29-1, p. 2-19, (476).

Describes the general geology and includes a table giving locations of collected specimens.

Burt, William A., and **Hubbard, Bela,** 1849, Reports on the linear surveys with reference to the mines and minerals in the northern peninsula of Michigan, SED 1, v. 3, 31-1, p. 802-935, 7 plates, 5 maps, (551).

Also in HED 5, v. 3, part 3, 31-1, p. 802-935, 7 plates, 5 maps, (571). In Jackson, Charles T., 1849.

Caine, Thomas A., 1903, Soil survey of the Hickory area, North Carolina, HD 473, v. 106, 57-2, p. 239-258, 2 maps, 1 plate, (4545).

Caine, Thomas A., 1904, Soil survey of the Fargo area, North Dakota, HD 746, v. 111, 58-2, p. 979-1003, 1 map (in v. 112), (4737-4738).

Caine, Thomas A., and **Bennett, Hugh H.,** 1905, Soil survey of Appomattox County, Virginia, HD 458, v. 108, 58-3, p. 151-168, 1 map (in v. 109), (4887-4888).

Caine, Thomas A., and **Kocher, A. E.,** 1904a, Soil survey of the Jamestown area, North Dakota, HD 746, v. 111, 58-2, p. 1005-1026, 1 map (in v. 112), (4737-4738).

Caine, Thomas A., and **Kocher, A. E.,** 1904b, Soil survey of the Paris area, Texas, HD 746, v. 111, 58-2, p. 533-562, 1 map (in v. 112), (4737-4738).

Caine, Thomas A., and **Lyman, W. S.,** 1905a, Soil survey of the San Antonio area, Texas, HD 458, v. 108, 58-3, p. 447-473, 1 map (in v. 109), (4887-4888).

Caine, Thomas A., and **Lyman, W. S.,** 1905b, Soil survey of the Superior area, Wisconsin-Minnesota, HD 458, v. 108, 58-3, p. 751-768, 1 map (in v. 109), (4887-4888).

Caine, Thomas A., and **Lyman, W. S.,** 1905c, Soil survey of the Wooster area, Ohio, HD 458, v. 108, 58-3, p. 543-564, 1 map (in v. 109), (4887-4888).

Caine, Thomas A., and **Mangum, A. W.,** 1903, Soil survey of the Mount Mitchell area, North Carolina, HD 473, v. 106, 57-2, p. 259-271, 1 map, 3 plates, (4545).

Caine, Thomas A., and **Schroeder, Frank C.,** 1908, Soil survey of Montgomery County, Mississippi, HD 352, v. 75, 59-2, p. 385-404, 1 map (in v. 76), (5178-5179).

Caine, Thomas A., and **Tailby, G. W., Jr.,** 1908, Soil survey of the Wheeling area, West Virginia, HD 352, v. 75, 59-2, p. 167-194, 1 map (in v. 76), (5178-5179).

Cameron, Frank K., 1900, Application of the theory of solutions to the study of soils, HD 399, v. 88, 56-1, p. 141-172, (3985).

Describes work done under the direction of the Soil Survey.

Cameron, Frank K., 1901, Application of the theory of solution to the study of soils, HD 526, v. 107, 56-2, p. 423-453, (4181).

Describes work done under the direction of the Soil Survey.

Campbell, Donald, 1874, Geological observations made between Pueblo and Fort Garland, Colorado Territory, HED 193, v. 12, 43-1, p. 59-61, (1610).

In Ruffner, E. H., 1874.

Campbell, John V., 1871, Earthquake in Peru, August 13, 1868, HED 20, v. 2, 42-1, p. 421-425, (1471).

An eyewitness description of the earthquake and its results. In the Annual Report of the Smithsonian Institution.

Canudas, A., 1859, Record of earthquakes felt at the Collegiate Seminary of Guatemala in 1857 and 1858, SMD 49, v. 1, 35-2, p. 437, (993).

Also in HMD 57, v. 1, 35-2, p. 437, (1016).

Carpenter, Phillip P., 1861, Lectures on Mollusca or "shell-fish" and their allies, SMD 21, v. 1, 36-2, p. 151-283, (1089).

In the Annual Report of the Smithsonian Institution. Includes descriptions and discussions of fossil forms.

Carr, E. P., and **Mangum, A. W.,** 1907a, Soil survey of the Everett area, Washington, HD 925, v. 110, 59-1, p. 1053-1079, 1 map (in v. 111), (5050-5051).

Carr, E. P., and **Mangum, A. W.,** 1907b, Soil survey of Island County, Washington, HD 925, v. 110, 59-1, p. 1033-1051, 1 map (in v. 111), (5050-5051).

Carr, E. P., et al, 1907, Soil survey of Dallas County, Alabama, HD 925, v. 110, 59-1, p. 453-472, 1 map (in v. 111), (5050-5051).

Carr, M. Earl, and **Belden, H. L.,** 1905, Soil survey of Saline County, Missouri, HD 458, v. 108, 58-3, p. 791-814, 1 map (in v. 109), (4887-4888).

Carr, M. Earl, and **Bennett, Frank,** 1907, Soil survey of Henderson County, Tennessee, HD 925, v. 110, 59-1, p. 643-657, 1 map (in v. 111), (5050-5051).

Carr, M. Earl, Griffin, A. M., and **Lee, Ora, Jr.,** 1908, Soil survey of Madison County, New York, HD 352, v. 75, 59-2, p. 119-165, 1 map (in v. 76), (5178-5179).

Carr, M. Earl, and **Tharp, W. E.,** 1908, Soil survey of the Waycross area, Georgia, HD 352, v. 75, 59-2, p. 303-333, 1 map (in v. 76), (5178-5179).

Carson, J. Petigru, 1874, Geological report upon the Darien Route and Nercalagua River, Bay of San Blas, HMD 113, v. 5, 42-3, p. 127-140, 2 plates, (1575).

Describes the geology of parts of Panama. In Selfridge, Thomas O., 1874.

Carter, William T., Jr., and **Kocher A. E.,** 1905, Soil survey of Anderson County, Texas, HD 458, v. 108, 58-3, p. 397-420, 1 map (in v. 109), (4887-4888).

Carter, William T., Jr., and **Kocher, A. E.,** 1907, Soil survey of Houston County, Texas, HD 925, v. 110, 59-1, p. 537-565, 1 map (in v. 111), 5050-5051).

Carter, William T., Jr., and **Lyman, W. S.,** 1904, Soil survey of the Leesburg area, Virginia, HD 746, v. 111, 58-2, p. 191-231, 1 map (in v. 112), (4737-4738).

Carter, William T., Jr., and Smith Howard C., 1908, Soil survey of Riley County, Kansas, HD 352, v. 75, 59-2, p. 911-941, 1 map (in v. 76), (5178-5179).

Carter, William T., Jr., et al, 1908, Soil survey of Prarie County, Arkansas, HD 352, v. 75, 59-2, p. 629-660, 1 map (in v. 76), (5178-5179).

Cass, Lewis, 1825, Letter . . . on the advantage of purchasing the country upon Lake Superior where copper had been found, SD 19, v. 2, 18-2, p. 3-4, (109).

Chamberlain, P. W., 1903, The volcanoes of Nicaragua, SD 131, v. 9, 57-2, p. 27-33, (4424).

See also Jones, James O., 1903.

Chamberlain, T. C., 1901, On Lord Kelvin's address on the age of the earth as an abode fitted for life, HD 737, part 1, v. 119, 56-1, p. 223-246, (4016).

Discusses the age of the earth from a standpoint of uniformitarianism and argues against Kelvin's age determination based on heat loss. In the Annual Report of the Smithsonian Institution.

Chandler, M. T. W., 1857, (Report upon the area) from San Vincento to Presidio del Norte, SED 108, v. 20, part 1, 34-1, p. 80-85, (832).

Also in HED 135, v. 14, part 1, 34-1, p. 80-85, (861). Part of the Mexican Boundary Survey Reports. In Emory, William H., 1857.

Chaney, L. W., 1905, Glacial exploration in the Montana Rockies, HD 460, v. 111, 58-3, p. 493-496, (4890).

In the Report of the 8th International Geographic Congress.

Channing, William F., 1848a, (Report of explorations on the St. Marys River), SED 2, v. 2, 30-1, p. 199-206, (504).

Describes the geology of parts of Upper Michigan. In Jackson, Charles T., 1848a.

Channing, William F., 1848b, Synopsis of a survey of St. Marys River, SED 2, v. 2, 30-1, p. 209, (504).

A progress report on the geological exploration of parts of Upper Michigan. In Jackson, Charles T., 1848a.

Channing, William F., 1848c, Synopsis of (a survey of part of Upper Michigan), SED 2, v. 2, 30-1, p. 207-208, (504).

A progress report on the geological exploration of parts of Upper Michigan. In Jackson, Charles T., 1848a.

Chevalier, A., 1905, Récentes explorations scientifiques dans L'intérier du Soudan, HD 460, v. 111, 58-3, p. 690-695, (4890).

Describes the topography and general geology of the area. In the Report of the 8th International Geographic Congress.

Chree, C., 1893, Some applications of physics and mathematics to geology, HMD 334, part 1, v. 43, 52-1, p. 127-153, (3001).

A discussion of the shape, structure, and elasticity of the earth. In the Annual Report of the Smithsonian Institution.

Christen, T., 1905, Zur Dynamik der Sinkstoffe, HD 460, v. 111, 58-3, p. 523-530, (4890).

A discussion of the hydrodynamics of ground water. In the Report of the 8th International Geographic Congress.

Church, John A., 1877, Preliminary report on examination at the Comstock Lode, HED 1, part 2, v. 4, 45-2, p. 1284-1285, (1796).

Describes the economic geology of the area. In Wheeler, George M., 1877.

Church, John A., 1878, Report upon examination of the Comstock Lode, HED 1, part 2, v. 5, 45-3, p. 1567-1588, (1846).

Describes the mining and economic geology of the area. In Wheeler, George M., 1878.

Clarke, F. W., 1889, The meteorite collection in the U. S. National Museum, SMD 170, part 2, v. 12, 49-2, p. 225-265, (2499).

In the Annual Report of the U. S. National Museum.

Clover, Richardson, 1892, Report of the results of the survey for the purpose of determining the practicability of laying a telegraphic cable between the United States and the Hawaiian Islands, SED 153, v. 2, 52-1, p. 1-28, 6 plates, 7 maps and charts, (2893).

Describes the known bottom features and sediments of parts of the Pacific Ocean.

Coffee, George N., and **Hearn, W. Edward,** 1902a, soil survey of Alamance County, North Carolina, HD 655, v. 117, 57-1, p. 297-310, 1 map, 3 plates, (4384).

Coffey, George N., and **Hearn, W. Edward,** 1902b, Soil survey of the Cary area, North Carolina, HD 655, v. 117, 57-1, p. 311-315, 2 maps, (4384).

Coffey, George N., et al, 1903a, Soil survey of Clay County, Illinois, HD 473, v. 106, 57-2, p. 533-548, 1 map, 1 plate, (4545).

Coffey, George N., et al, 1903b, Soil survey of St. Clair County, Illinois, HD 473, v. 106, 57-2, p. 507-532, 1 map, 2 plates, (4545).

Coffey, George N., et al, 1904a, Soil survey of Johnson County, Illinois, HD 746, v. 111, 58-2, p. 721-736, 1 map (in v. 112), (4737-4738).

Coffey, George N., et al, 1904b, Soil survey of Knox County, Illinois, HD 746, v. 111, 58-2, p. 737-752, 1 map (in v. 112), (4737-4738).

Coffey, George N., et al, 1904c, Soil survey of McLean County, Illinois, HD 746, v. 111, 58-2, p. 777-797, 1 map (in v. 112), (4737-4738).

Coffey, George N., et al, 1904d, Soil survey of Sangamon County, Illinois, HD 746, v. 111, 58-2, p. 703-719, 1 map (in v. 112), (4737-4738).

Coffey, George N., et al, 1904e, Soil survey of Winnebago County, Illinois, HD 746, v. 111, 58-2, p. 753-775, 1 map (in v. 112), (4737-4738).

Coffin, Frederick B., 1892, Special report of work in the artesian and underflow investigation . . . , in South Dakota, SED 41, part 4, v. 4, 52-1, p. 51-61, 2 plates, (2899).

Describes the areal geology and the ground water conditions for parts of South Dakota.

Coleman, A. P., 1905, Glacial lakes and Pleistocene changes in the St. Lawrence Valley, HD 460, v. 111, 58-3, p. 480-486, (4890).

In the Report of the 8th International Geographic Congress.

Collins, Frederick, 1879, Report on a survey of the proposed route for an interoceanic canal by way of the Atrato, Napipi, and Doguado Rivers in the . . . United States of Colombia, SED 75, v. 2, 45-3, p. 59-124, 9 maps and profiles, (1829).

Describes the landforms of parts of Panama and Colombia. Bound in the same volume as SED 113, the Annual Report of the Coast and Geodetic Survey.

Collins, P. McD., 1858, Explorations of Amoor River, HED 98, v. 12, 35-1, p. 1-67, 3 maps, (958).

Describes travels through parts of Siberia, Manchuria, and China. It includes general landform and lithologic descriptions, and a description of the known mineral resources of Siberia.

Comstock, G. B., 1876, Notes on European Surveys, HED 1, part 2, v. 5, 44-2, p. 126-217, 8 maps, (1745).

Describes the organization and work of the Geological and Topographical Surveys in Austria, Belgium, England, Germany, Italy, Norway, Russia, Spain, Sweden, and Switzerland.

Comstock, Theodore B., 1874, Geological report (upon northern Wyoming), HED 285, v. 17, 43-1, p. 85-184, 4 plates, 1 map, (1615).

In Jones, William A., 1874.

Congres Geologique International, 1893, Compte rendue de la 5me session, Washington 1891, HMD 107, v. 13, 53-2, p. 1-529, 22 maps and plates, (3241).

See also Emmons, S. F., 1893; McGee, W. J., et al, 1893; and Van Hise, C. R., 1893.

Conkling, A. R., 1876, Report on the geology of the mountain ranges from La Veta Pass to the head of the Pecos, HED 1, part 2, v. 5, 44-2, p. 419-422, (1745).

Describes the general geology of parts of New Mexico. In Wheeler, George M., 1876.

Conkling, A. R., 1877a, Geologic report on the portions of western Nevada and eastern California between the parallels of 39° 30' and 38° 30' . . . , HED 1, part 2, v. 4, 45-2, p. 1285-1295, (1796).

In Wheeler, George M., 1877.

Conkling, A. R., 1877b, Report on the foot hills facing the plains from latitude 35° 30' to 38° approximately, HED 1, part 2, v. 4, 45-2, p. 1298-1303, (1796).

Describes the general geology of parts of Colorado and New Mexico. In Wheeler, George M., 1877.

Conkling, A. R., 1877c, Report on the lithology of portions of southern Colorado, and northern New Mexico, HED 1, part 2, v. 4, 45-2, p. 1295-1298, (1796).

In Wheeler, George M., 1877.

Conkling, A. R., 1878, Geologic report on portions of western Nevada and eastern California . . . , HED 1, part 2, v. 5, 45-3, p. 1589-1607, (1846).

In Wheeler, George M., 1878.

Conrad, T. A., 1855a, Description of the fossil shells (found while exploring routes in California), SED 78, v. 13, part 5, 33-2, p. 317-329, 9 plates, (762).

Also in HED 91, v. 11, part 5, 33-2, p. 317-329, 9 plates, (795). Part of the Pacific Railroad Survey Reports. In Williamson, R. S., 1855.

Conrad, T. A., 1855b, Description of the Tertiary fossils collected on the survey (from Sacramento Valley to the Columbia River), SED 78, v. 11, part 6, 33-2, p. 69-73, 4 plates, (763).

Also in HED 91, v. 11, part 6, 33-2, p. 69-73, 4 plates, (796). Part of the Pacific Railroad Survey Reports. In Newberry, J. S., 1855.

Conrad, T. A., 1855c, Remarks on the fossil shells from Chile, HED 121, v. 15, part 2, 33-1, p. 282-286, 2 plates, (729).

Describes Cretaceous, Tertiary, and Recent shells. In volume 2 of the U.S. Astronomical Expedition to the Southern Hemisphere. See also Gilliss, J. M., 1855.

Conrad, T. A., 1855d, Report on the paleontology of the survey (of routes in California), SED 78, v. 13, part 7, 33-2, p. 189-196, 10 plates, (764).

Also in HED 91, v. 11, part 7, 33-2, p. 189-196, 10 plates, (797). Part of the Pacific Railroad Survey Reports. In Antisell, Thomas, 1855.

Conrad, T. A., 1857, Descriptions of Cretaceous and Tertiary fossils (found during the United States and Mexican Boundary Survey), SED 108, v. 20, part 1, 34-1, p. 141-174, 20 plates, (832).

Also in HED 135, v. 14, part 1, 34-1, p. 141-174, 20 plates, (861). In Parry, C. C., and Schott, Arthur, 1857.

Cook, Frederick A., 1905, Results of a journey around Mount McKinley, HD 460, v. 111, 58-3, p. 758-762, (4890).

Describes the topography of the area. In the Report of the 8th International Geographic Congress.

Cope, E. D., 1872a, On the fishes of the Tertiary shales of Green River, Wyoming Territory, HED 325, v. 15, 42-2, p. 425-431, (1520).

In Hayden, F. V., 1872b.

Cope, E. D., 1872b, On the fossil reptiles and fishes of the Cretaceous rocks of Kansas, HED 325, v. 15, 42-2, p. 385-424, (1520).

In Hayden, F. V., 1872b.

Cope, E. D., 1872c, On the geology and paleontology of the Cretaceous strata of Kansas, HED 326, v. 15, 42-2, p. 318-349, (1520).

In Hayden, F. V., 1872a.

Cope, E. D., 1872d, On the vertebrate fossils of the Wahsatch strata, HED 326, v. 15, 42-2, p. 350-353, (1520).

In Hayden, F. V., 1872a.

Cope, E. D., 1873, On the extinct vertebrata of Wyoming, with notes on the geology, HMD 112, v. 3, 42-3, p. 545-649, 6 plates, (1573).

In Hayden, F. V., 1873.

Cope, E. D., 1874, Notes on the Eocene and Pliocene lacustrine formations of New Mexico . . ., HED 1, part 2, v. 4, 43-2, p. 591-606, (1637).

Describes the Tertiary mammalia from New Mexico. Accompanies Wheeler, George M., 1874.

Cope, E. D., 1875, Report on the geology of that part of northwestern New Mexico examined during the field season of 1874, HED 1, part 2, v. 5, 44-1, p. 981-1017, (1676).

Describes the invertebrate and vertebrate paleontology of the area. In Wheeler, George M., 1875.

Cope, E. D., 1884, The vertebrata of the Tertiary formations of the west, HMD 60, v. 27, 48-1, p. 1-1009, 134 plates, (2239).

This is the 3rd volume of Hayden's final report and is the only part published as a congressional document.

Cope, E. D., 1885, The structure of the columella auris in the Pelycosauria, SMD 69, v. 5, 48-1, p. 93-95, (2269).

In the Memoirs of the National Academy of Sciences (v. 3, part 1, memoir 7).

Cope, E. D., 1886, On two new forms of polydont and gonorhynchid fishes from the Eocene of the Rocky Mountains, SMD 154, v. 7, 49-1, p. 161-165, 1 plate, (2348).

In the Memoirs of the National Academy of Sciences (v. 3, part 2, memoir 17).

Cowden, John, 1878, Memorial on the subject of the improvement of the navigation of the Mississippi River and the reclamation of the low lands of the Mississippi Valley, SMD 89, v. 2, 45-2, p. 1-11, (1786).

Describes the fluvial features and floods of the Mississippi Valley.

Craven, T. A., 1855, Description of a specimen box for bringing up the bottom in deep-sea soundings . . ., SED 10, v. 12, 33-2, p. 191-192, 1 plate, (757).

Describes an early sampling device used at depths of from 300 to 2400 feet. In the Annual Report of the Coast Survey.

Creak, Etterick W., 1904, Terrestrial magnetism in its relation to geography, HD 748, v. 114, 58-2, p. 391-406, 2 maps, (4740).

In the Annual Report of the Smithsonian Institution.

Crookes, William, 1898, Diamonds, HD 575, part 1, v. 78, 55-2, p. 219-235, (3706).

Describes the origin, properties, and known deposits of diamonds. In the Annual Report of the Smithsonian Institution.

Cross, Osborne, 1850, Journal of the march of the regiment of mounted rifleman to Oregon, from May 10 to October 5, 1849, SED 1, v. 1, part 2, 31-2, p. 127-244, 35 plates, (587).

Describes the general topography and landforms.

Culbertson, Thaddeus A., 1851, Journal of an expedition to the Mauvaises Terres and the Upper Missouri in 1850, SMD 1, v. 1, 32-special, p. 84-145 (607).

Describes the general topography and lithology.

Culver, Garry E., 1892, Report (on the ground water of South Dakota), SED 41, part 3, v. 4, 52-1, p. 191-209, 3 plates and maps, (2899).

Describes the areal geology and the ground water conditions of parts of South Dakota.

Cuvier, Georges, 1861, Memoir of Rene-Just Häuy, SMD 21, v. 1, 36-2, p. 376-392, (1089).

Describes the life and work of this famous mineralogist. In the Annual Report of the Smithsonian Institution.

Dana, Edward S., 1884, Record of recent scientific progress in Mineralogy, HMD 26, v. 5, 47-2, p. 533-549, (2121).

In the Annual Report of the Smithsonian Institution.

Dana, Edward S., 1885a, Record of scientific progress in Mineralogy for 1883, HMD 69, v. 34, 48-1, p. 661-679, (2246).

In the Annual Report of the Smithsonian Institution.

Dana, Edward S., 1885b, Record of scientific progress, 1884 in Mineralogy, SMD 33, part 1, v. 2, 48-2, p. 543-561, (2266).

In the Annual Report of the Smithsonian Institution.

Dana, Edward S., 1886, Record of scientific progress, 1885 in Mineralogy, HMD 15, v. 25, 49-1, p. 687-712, (2431).

Includes a bibliography for 1885. In the Annual Report of the Smithsonian Institution.

Dana, Edward S., 1889, Mineralogy in 1886, HMD 600, part 1, v. 17, 50-1, p. 449-476, (2581).

Includes a bibliography for 1886. In the Annual Report of the Smithsonian Institution.

Dana, Edward S., 1890, Mineralogy for 1887 and 1888, HMD 142, part 1, v. 14, 50-2, p. 455-473, (2668).

In the Annual Report of the Smithsonian Institution.

Dana, Edward S., and **Grinnell, George B.,** 1876, Geological report (of a reconnaissance from Carroll, Montana, to Yellowstone National Park), HED 1, part 2, v. 5, 44-2, p. 657-694, (1745).

Describes the general geology. In Ludlow, William, 1876.

Dana, James D., 1882, Review of "The geology of the high plateaus of Utah" by C. E. Dutton, HED 1, part 2, v. 6, 47-1, p. 173-177, (2014).

Reviews a publication resulting from War Department explorations. See also Geikie, Archibald, 1882.

Dana, James D., 1890, A memoir of Asa Gray, HMD 142, part 1, v. 14, 50-2, p. 745-762, (2668).

Describes the life and work of this paleobotanist. In the Annual Report of the Smithsonian Institution.

Darton, Nelson H., 1899, North American geology for 1886, HMD 600, part 1, v. 17, 50-1, p. 189-229, (2581).

In the Annual Report of the Smithsonian Institution.

Darton, Nelson H., 1906, Geology of the Owl Creek Mountains, with notes on the resources of adjoining regions in the ceded portion of the Shoshone Indian Reservation, Wyoming, SD 219, v. 5, 59-1, p. 1-48, 15 plates, 4 maps, (4913).

Describes the general geology and the limestone resources of this area.

Daubree, A., 1894, Deep-sea deposits, HMD 184, part 1, v. 29, 53-2, p. 545-566, 2 maps, (3257).

Describes the marginal and abyssal sediments in light of data from the Challenger expedition. In the Annual Report of the Smithsonian Institution.

Daubree, G. A., 1862, Synthetical studies and experiments on metamorphism and on the formation of crystalline rocks, HMD 77, v. 1, 37-2, p. 228-304, (1141).

Discusses in detail experimental metamorphism, contact and regional metamorphism, mineral deposits and metamorphism, and the causes of metamorphism. In the Annual Report of the Smithsonian Institution.

Daubree, G. A., 1869, Synthetic experiments relative to meteorites—approximations to which these experiments lead, HED 83, v. 12, 40-3, p. 312-341, (1380).

Describes the ccmposition and discusses the origin of meteorites. In the Annual Report of the Smithsonian Institution.

Davidson, George, 1868, Report . . . on the resources and the coast features of Alaska Territory, HED 275, v. 18, 40-2, p. 187-329, (1344).

Describes the coastal topography and the known mineral resources. In the Annual Report of the Coast Survey.

Davis, Arthur, 1903, Hydrography of the American Isthmus, SD 124, v. 11, 57-2, p. 507-630, 11 plates, 3 maps, (4426).

Describes the hydrogeology and fluvial features of parts of Panama and Nicaragua.

Davis, C. H., 1866, Report . . . on interoceanic canals and railroads between the waters of the Atlantic and Pacific Oceans, SED 62, v. 21, 39-1, p. 2-19, 13 maps, (1238).

Describes the topography of various possible routes in Central America.

Davis, George W., et al, 1906, Report of the Board of Consulting Engineers for the Panama Canal, SD 231, v. 15, p. 1-99, (4923).

Describes the geology and the geologic engineering problems to be met during canal construction.

Davis, Jefferson, 1855, Report of the Secretary of War on the several railroad explorations, SED 78, v. 13, part 1, 33-2, p. 3-33, (758).

Also in HED 129, v. 18, part 1, 33-1, p. 3-43, (736); *and* HED 91, v. 11, part 1, 33-2, p. 3-33, (791). Part of the Pacific Railroad Survey Reports, this describes in a general fashion the various routes explored.

Davis, W. M., 1904, Complications of the geographical cycle, HD 460, v. 111, 58-3, p. 150-163, (4890).

In the Report of the 8th International Geographic Congress.

Davis, W. M., 1905a, Bearing of physiography upon Suess's theories, HD 460, v. 111, 58-3, p. 164, (4890).

An abstract in the Report of the 8th International Geographic Congress.

Davis, W. M., 1905b, College entrance examinations in physiography, HD 460, v. 111, 58-3, p. 956-957, (4890).

An abstract in the Report of the 8th International Geographic Congress.

Delafield, Richard, 1829, Survey of the passes at the mouth of the Mississippi, La., HED 7, v. 1, 21-1, p. 7-11, (195).

Describes the channel morphology and includes suggestions for the removal of bars.

de Martonne, E., 1905, Évolution morphologique des Karpates méridionales. HD 460, v. 111, 58-3, p. 138-145, (4890).

Describes landscape formation in parts of Hungary and Romania. In the Report of the 8th International Geographic Congress.

deMello, Carlos, 1905, Physiography and map drawing, HD 460, v. 111, 58-3, p. 239-243, (4890).

In the Report of the 8th International Geographic Congress.

de Quatrefages, J. L. A., 1863, Memoir of M. Isodore Geoffroy Saint Hilaire, HMD 25, 37-3, p. 384-394, (1172).

> An extensive review of Saint Hilaire's life and work. In the Annual Report of the Smithsonian Institution.

Derby, George H., 1850, Memoir on the Geology and Topography of California, SED 47, v. 10, 31-1, p. 3-16, 2 maps, (558).

Derby, George H., 1852a, Report of the expedition of the United States transport "Invincible" . . . to the Gulf of California, and the River Colorado, during . . . 1850 and 1851, SED 81, v. 9, 32-1, p. 2-28, 1 map, (620).

> Describes the hydrologic and topographic features of the lower Colorado River.

Derby, George H., 1852b, Report on a reconnoissance of the Tulare Valley, SED 110, v. 10, 32-1, p. 4-17, 1 map, (621).

> Describes the topography and lithology of parts of the California Coast Ranges and the San Joaquin Valley.

Derby, Orville A., 1907, The geology of the diamond and carbonado washings of Bahia, Brazil, HD 575, v. 97, 59-2, p. 215-222, 2 plates, (5200).

> In the Annual Report of the Smithsonian Institution.

Desor, E., 1851, On the superficial deposits of (the Lake Superior) district, SED 4, v. 3, 32-special, p. 232-273, (609).

> Describes the drift, terraces, and alluvial deposits of the area and ascribes their origin to processes other than glaciation. Forms chapter 14 of Foster, J. W., and Whitney, J. D., 1851.

Dickenson, George J., 1849, Report of . . ., SED 1, v. 3, 31-1, p. 503-506, (551).

> Also in HED 5, v. 3, part 3, 31-1, p. 503-506, (571). Describes the mining and areal geology of parts of the Lake Superior region.

Diller, J. S., 1898 Crater Lake, Oregon, HD 575, part 1, v. 78, 55-2, p. 369-379, 16 plates, (3706).

> Describes the lacustrine and volcanic features. In the Annual Report of the Smithsonian Institution.

Doane, Gustavus C., 1871 Report upon the so-called Yellowstone expedition of 1870, SED 51, v. 1, 41-3, p. 1-38, (1440).

> Describes the topography of parts of Montana and Wyoming and describes the thermal springs and geysers of what is now Yellowstone National Park.

Dodge, G. M., 1868, Report on the construction of the Union Pacific Railroad, HED 331, v. 20, 40-2, p. 1-22, (1346).

Describes the topography and mineral resources of parts of the Great Plains and Rocky Mountains. See also Van Lennep, D., 1868.

Dodge, G. M., 1869, Annual report of the Union Pacific Railroad Company, to the Secretary of the Interior, SED 10, v. 1, 40-3, p. 3-25, (1360).

Describes the topography and mineral resources of parts of the Great Plains and Rocky Mountains.

Dornbach, L. M., 1855, Report upon an analytical examination of water and minerals from the hot springs in Des Chutes Valley, SED, 78, v. 13, part 6, 33-2, p. 74-78, (763).

Also in HED 91, v. 11, part 6, 33-2, p. 74-78, (796). A part of Pacific Railroad Survey Reports.

Dorsey, Clarence W., 1901, A soil survey around Lancaster, Pa., HD 526, v. 107, 56-2, p. 61-84, 4 plates, 1 map, (4181).

Dorsey, Clarence W., 1907, Agricultural Bulletin No. 3: Soil conditions in the Philippines, HD 2, v. 10, 59-2, p. 363-392, (5113).

Describes the various soils found in the Philippines.

Dorsey, Clarence W., and **Bonsteel, J. A.,** 1900, A soil survey in the Connecticut Valley, HD 399, v. 88, 56-1, p. 125-140, 7 plates, 1 map, (3985).

Dorsey, Clarence W., and **Bonsteel, J. A.,** 1901, Soil survey of Cecil County, Md., HD 526, v. 107, 56-2, p. 103-124, 1 map, (4181).

Dorsey, Clarence W., and **Coffey, George N.,** 1901, Soil survey of Montgomery County, Ohio, HD 526, v. 107, 56-2, p. 85-102, 3 plates, 1 map, (4181).

Dorsey, Clarence W., Mesmer, Louis, and **Caine, Thomas A.,** 1903, Soil survey from Arecibo to Ponce, Porto Rico, HD 473, v. 106, 57-2, p. 793-839, 4 plates, 1 map, (4545).

Dorsey, Clarence W., et al, 1902, Soil survey of the Statesville area, North Carolina, HD 655, v. 117, 57-1, p. 273-295, 1 map, (4384).

Drake, J. A., 1904, Soil survey of the Parsons area, Kansas, HD 746, v. 111, 58-2, p. 891-909, 1 map (in v. 112), (4737-4738).

Drake, J. A., and **Belden, H. L.,** 1907a, Soil survey of Cherokee Country, South Carolina, HD 925, v. 110, 59-1, p. 333-349, 1 map (in v. 111), 5050-5051).

Drake, J. A., and **Belden, H. L.,** 1907b, Soil survey of York County, South Carolina, HD 925, v. 110, 59-1, p. 309-332, 1 map (in v. 111), (5050-5051).

Drake, J. A., and **Belden, H. L.,** 1908, Soil survey of New Hanover County, North Carolina, HD 352, v. 75, 59-2, p. 245-279, 1 map (in v. 76), (5178-5179).

Drake, J. A., and **Strahorn, A. T.,** 1905, Soil survey of Webster County, Missouri, HD 458, v. 108, 58-3, p. 845-858, 1 map (in v. 109), (4887-4888).

Drake, J. A., and **Tharp, W. E.,** 1905, Soil survey of Allen County, Kansas, HD 458, 58-3, p. 875-894, 1 map (in v. 109), (4887-4888).

Drude, Oscar, 1905, Die Methode der Pflanzengeographischen Kartographie erläutert an der Flora von Sachsen, HD 460, v. 111, 58-3, p. 608-612, (4890).

Describes the Quaternary and Recent plant distribution of parts of Germany. In the Report of the 8th International Geographic Congress.

Dubois, Eugene, 1899, Pithecanthropus erectus—a form from the ancestral stock of mankind, HD 309, v. 91, 55-3, p. 445-459, 3 plates, (3833).

In the Annual Report of the Smithsonian Institution.

Dudley, Timothy, 1859, The earthquake of 1811 at New Madrid, Missouri, SMD 49, v. 1, 35-2, p. 421-424, (993).

Also in HMD 57, v. 1, 35-2, p. 421-424, (1016). Describes the topographic changes and general destruction of the earthquake. In the Annual Report of the Smithsonian Institution.

du Pre, Warren, 1875, On a series of earthquakes in North Carolina, commencing on the 10th of February, 1874, HMD 56, v. 2, 43-2, p. 254-260, (1654).

Describes the topographic changes caused by the earthquakes. In the Annual Report of the Smithsonian Institution.

Dutton, C. E., 1902, A general description of the volcanic phenomina found in that portion of Central America traversed by the Nicaragua Canal . . ., SD 357, v. 26, 57-1, p. 55-62, (4245).

See also Wheeler, E. S., 1902.

Eaton, A. K., Keyes, W. S., and **DeLacy, W. W.,** 1869, Notes on Montana, HED 54, v. 9, 40-3, p. 134-160, (1374).

Describes the mineral resources of parts of Montana.

Eckel, Edwin C., 1903, Portland cement manufacture, SD 19, v. 2, 58-1, p. 2-11, (4563).

Describes the limestone resources of parts of Alabama. See also Smith, Eugene A., 1903.

Eckel, Edwin C., and **Crider, A. F.,** 1905, Geology and cement resources of the Tombigbee River district, Mississippi-Alabama, SD 165, v. 4, 58-3, p. 1-23, 1 map, (4766).

Describes the general geology and the limestone resources of this area.

Edmunds, J. M., et al, 1863, (Letters concerning the) mineral resources of Nevada Territory, HED 26, v. 5, 37-3, p. 1-14, (1161).

These letters describe mineral deposits in various parts of Nevada. Note, there are 2 volumes labeled 5, HED 26 is in the second of these.

Egleston, T., 1873, Scheme for the qualitative determination of substances by the blowpipe, HMD 107, v. 3, 42-3, p. 172-184, (1573).

In the Annual Report of the Smithsonian Institution.

Eldridge, G. H., 1898a, The coast from Lynn Canal to Prince William Sound, HD 172, v. 13, 55-3, p. 103-104, (3737).

Describes the geology and coal deposits of part of Alaska.

Eldridge, G. H., 1898b, The Sushitna drainage area, HD 172, v. 13, 55-3, p. 111-112, (3737).

Describes the geology and economic deposits of this part of Alaska.

Eldridge, G. H., 1899, The extreme southeastern coast (of Alaska), HD 172, v. 13, 55-3, p. 100-102, (3737).

Describes the geology and economic deposits.

Eldridge, G. H., and **Muldrow, Robert,** 1899, Report of the Sushitna expedition, SD 172, v. 13, 55-3, p. 15-27, (3737).

Describes the geology and economic deposits of this part of Alaska.

Elie de Beaumont, Leonce, 1871, Memoir of Auguste Bravais, HED 153, v. 12, 41-3, p. 145-168, (1460).

Describes the life and work of this great mineralogist. In the Annual Report of the Smithsonian Institution.

Ellet, Charles, Jr., 1851, Report on the improvement of the navigation across the bars at the mouth of the Mississippi River, SED 17, v. 3, 31-2, p. 2-18, (589).

Describes the sedimentation, bars, and mud lumps in the passes of the Mississippi River.

Ellet, Charles, Jr., 1852, Report on the overflows of the delta of the Mississippi, SED 20, v. 4, 32-1, p. 13-106, 6 plates, (614).

Describes the hydrography, topography, and sedimentation on the Mississippi Delta.

Elliott, R. S., 1872, Report on the industrial resources of western Kansas and eastern Colorado, HED 325, v. 15, 42-2, p. 442-458, (1520).

A part of Hayden, F. V., 1872b.

Ely, Charles W., Coffey, George N., and **Griffen, A. M.,** 1905, Soil survey of Tama County, Iowa, HD 458, v. 108, 58-3, p. 769-790, 1 map (in v. 109), (4887-4888).

Ely, Charles W., and **Griffen, A. M.,** 1905, Soil survey of Dodge County, Georgia, HD 458, v. 108, 58-3, p. 231-246, 1 map (in v. 109), (4887-4888).

Ely, Charles W., and **Kocher, A. E.,** 1908, Soil survey of the Henderson area, Texas, HD 352, v. 75, 59-2, p. 459-480, 1 map (in v. 76), (5178-5179).

Ely, Charles W., Marean, Herbert W., and **Neill, N. P.,** 1907, Soil survey of East Baton Rouge Parish, Louisiana, HD 925, v. 110, 59-1, p. 517-535, 1 map (in v. 111), (5050-5051).

Ely, Charles W., Willard, Rex E., and **Weaver, J. T.,** 1908, Soil survey of Ransom County, North Dakota, HD 352, v. 75, 59-2, p. 963-997, 1 map (in v. 76), (5178-5179).

Emerson, Benjamin K., 1879, On the geology of Frobisher Bay and Field Bay, SED 27, v. 3, 45-3, p. 553-583, (1830).

Describes the petrology and paleontology of parts of the Arctic and of the Northwest Territories.

Emmons, S. F., 1893, Geological guidebook of the Rocky Mountain excursion, HMD 107, v. 13, 53-2, p. 253-487, 14 plates and maps, (3241).

A field guide of a trip conducted by the 5th International Geologic Congress. See also Congrès Geologique International, 1893.

Emmons, S. F., 1898, Map of Alaska . . . with descriptive text containing sketches of the geography, geology, and gold deposits and routes to the gold fields, SD 195, v. 11, 55-2, p. 1-44, 1 map, (3600).

This document, intended to aid prospectors, describes the areal geology and the gold deposits.

Emmons, S. F., 1905, Theories of ore deposition historically considered, HD 430, v. 106, 58-3, p. 309-336, (4885).

>Discusses the origin of ores and the history of ideas regarding the origin of ores. In the Annual Report of the Smithsonian Institution.

Emory, William H., 1848, Notes of a military reconnaissance from Fort Leavenworth in Missouri, to San Diego in California, SED 7, v. 3, 30-1, p. 1-416, 43 plates, 1 map, (505).

>Also in HED 41, v. 4, 30-1, p. 1-614, 67 plates, 3 maps. Includes descriptions of the topography and lithology along the route. See also Abert, J. W., 1848.

Emory, William H., 1855, Extract from report of a military reconnaissance made in 1846 and 1847, SED 78, v. 13, part 2, 33-2, p. 1-22, (759).

>Also in HED 91, v. 11, part 2, p. 1-22, (759). Published with the Pacific Railroad Survey Reports, but not a result of these surveys.

Emory, William H., 1857, Report on the United States and Mexican Boundary Survey, SED 108, v. 20, part 1, 34-1, p. 1-258, 88 plates, 1 map, (832).

>Also in HED 135, v. 14, part 1, 34-1, p. 1-258, 88 plates, 1 map, (861). Describes in detail the topography and geographic features of the boundary area.

Endlich, Frederic M., 1874, Report . . . upon lithological and geognostic specimens collected in Nicaragua, SED 57, v. 3, 43-1, p. 129, (1582).

>See also Hatfield, Chester, 1874.

Engelmann, Henry, 1858, Report of a geological exploration from Fort Levenworth to Bryan's Pass . . ., SED 11, v. 3, 35-1, p. 489-517, (920).

>Also in HED 2, v. 2, part 2, 35-1, p. 489-517, (943). Describes the stratigraphy and lithology of the route in detail and also describes the general geology of the Black Hills and of the Medicine Bow Mountains. See also Shumard, B. F., 1858.

Engelmann, Henry, 1859, . . . geology of the country between Fort Bridger and Camp Floyd, Utah Territory . . . , SED 40, v. 10, 35-2, p. 45-75, (984).

>Describes the hot springs, igneous rocks, and general geology of the area.

Engle, F., 1860, (Preliminary report of the Chiriqui Commission), SED 1, v. 3, part 1, 36-2, p. 36-38, (1080).

>Describes the topography, coal deposits, and gold deposits of parts of Panama. See also Evans, John, 1860.

Evans, John, 1860, (Preliminary report on Coal discovered by the Chiriqui Commission), SED 1, v. 3, part 1, 36-2, p. 43-44, (1080).

Describes coal beds discovered in Panama. See also Engle, F., 1860.

Eveland, Arthur J., 1907, Mining Bulletin No. 4: Preliminary reconnoissance of the Mancayan-Suyoc mineral region, Lepanto, Luzon, P. I., HD 2, v. 10, 59-2, p. 751-778, (5113).

Describes the areal geology and mineral deposits of this part of the Philippines.

Fahs, C. F., 1856, Report of an exploration of Peel Island, HED 97, v. 12, part 2, 33-2, p. 75-78, (803).

Also in SED 79, v. 14, part 2, 33-2, p. 75-78, (770). In volume 2 of the report of the Perry expedition to Japan and China. See also Perry, M. C., 1856.

Farlow, William G., 1890, Memoir of Asa Gray, HMD 142, part 1, v. 14, 50-2, p. 763-783, (2668).

Describes the life and work of this Paleobotanist. In the Annual Report of the Smithsonian Institution.

Favre, Ernest, 1879, Louis Agassiz, a biographical sketch, SMD 59, v. 3, 45-3, p. 236-261, (1835).

Describes the life and work of Agassiz. In the Annual Report of the Smithsonian Institution.

Featherstonhaugh, George W., 1833, (Letter asking for aid and giving reasons for a geological reconnaissance of the United States), SED 35, v. 1, 22-2, p. 5-7, (230).

An application for financial aid to support his proposed *Geological Journal*.

Featherstonhaugh, George W., 1835, Geological report of . . . the elevated country between the Missouri and Red Rivers, HED 151, v. 4, 23-2, p. 1-97, 1 plate, (274).

Also in SED 153, v. 4, 23-2, p. 2-43, (269). One of the earliest detailed geological descriptions of the country. The structure section mentioned in the text is not found in some copies of the House Document.

Featherstonhaugh, George W., 1836, Report of a geological reconnaissance . . . from Green Bay and Wisconsin Territory to the Coteau de Prairie . . . SED 333, v. 4, 24-1, p. 1-168, 3 plates, (282).

The second of Featherstonhaugh's detailed reconnaissances. It includes a correlation of American strata with the European stratigraphic section.

Fergusson, D., 1863, Report . . . on the country, its resources, and the route between Tucson and Lobos Bay, SED 1, v. 1, 37-special, p. 2-22, 3 maps, (1174).

Desribes the topography and mineral resources of parts of Arizona and Sonora.

Fippin, Elmer O., 1903a, Soil survey of the Dubuque area, Iowa, HD 473, v. 106, 57-2, p. 571-592, 1 map, (4545).

Fippin, Elmer O., 1903b, Soil survey of Howell County, Missouri, HD 473, v. 106, 57-2, p. 593-609, 2 plates, 1 map, (4545).

Fippin, Elmer O., 1904, Soil survey of the Connecticut Valley, HD 746, v. 111, 58-2, p. 39-61, 2 maps (in v. 112), (4737-4738).

Fippin, Elmer O., and **Burgess, James L.,** 1905, Soil survey of the Cando area, North Dakota, HD 458, v. 108, 58-3, p. 925-949, 1 map (in v. 109), (4887-4888).

Fippin, Elmer O., and **Carter, William T., Jr.,** 1907, Soil survey of the Binghamton area, New York, HD 925, v. 110, 59-1, p. 71-96, 1 map (in v. 111), (5050-5051).

Fippin, Elmer O., and **Drake, J. A.,** 1905a, Soil survey of the Bainbridge area, Georgia, HD 458, v. 108, 58-3, p. 247-267, 1 map (in v. 109), (4887-4888).

Fippin, Elmer O., and **Drake, J. A.,** 1905b, Soil survey of the O'Fallon area, Missouri-Illinois, HD 458, v. 108, 58-3, p. 815-843, 1 map (in v. 109), (4887-4888).

Fippin, Elmer O., and **Mann, C. W.,** 1908, Soil survey of Niagara County, New York, HD 352, v. 75, 59-2, p. 69-117, 1 map (in v. 76), (5178-5179).

Fippin, Elmer O., and **Rice, Thomas D.,** 1902, Soil survey of Allegan County, Michigan, HD 655, v. 117, 57-1, p. 93-124, 9 plates, 2 maps, (4384).

Fippin, Elmer O., and **Root, Aldert S.,** 1904, Soil survey of Gadsden County, Florida, HD 746, v. 111, 58-2, p. 331-353, 1 map (in v. 112), (4737-4738).

Fischer, P., 1873, The scientific labors of Edward Lartet, HMD 107, v. 3, 42-3, p. 172-184, (1573).

Describes the life and work of this paleontologist. In the Annual Report of the Smithsonian Institution.

Fischer, Theobald, 1905, Morocco, HD 430, v. 106, 58-3, p. 355-372, (4885)

Describes the landforms and general geology of the country. In the Annual Report of the Smithsonian Institution.

Fiske, John, 1901, Reminiscences of Huxley, HD 537, part 1, v. 114, 56-2, p. 713-728, (4188).

Describes the life and work of T. H. Huxley. In the Annual Report of the Smithsonian Institution.

Flourens, P., 1862, Memoir of Geoffroy Saint Hilaire, HMD 77, v. 1, 37-2, p. 161-174, (1141).

Describes the life and work of this French paleontologist. In the Annual Report of the Smithsonian Institution.

Flourens, P., 1863, Memoir of Leopold von Buch, HMD 25, v. 2, 37-3, p. 358-372, (1172).

Describes the life and work of this great geologist. In the Annual Report of the Smithsonian Institution.

Flourens, P., 1866, Memoir of Ducrotay de Blainville, HED 102, v. 14, 39-1, p. 175-188, (1265).

Describes the life and work of this French paleontologist. In the Annual Report of the Smithsonian Institution.

Flourens, P., 1869a, History of the works of Cuvier, HED 83, v. 12, 40-3, p. 141-165, (1380).

Reviews the scientific work of Cuvier. See also Flourens, P., 1869b. In the Annual Report of the Smithsonian Institution.

Flourens, P., 1869b, Memoir of Cuvier, HED 83, v. 12, 40-3, p. 121-140, (1380).

Describes the life and work of Cuvier. See also Flourens, P., 1869a. In the Annual Report of the Smithsonian Institution.

Forshey, C. G., 1875, Report of survey and borings made at the proposed site of the Lake Borgne outlet, HED 127, v. 15, 43-2, p. 95-104, (1648).

Describes the sediments and the fluvial and lacustrine features of part of the Mississippi Delta. See also Warren, G. K., 1875a.

Foster, J. W., 1848, (Synopsis of a report on the geology of part of the Lake Superior region), SED 2, v. 2, 30-2, p. 159-163, (530).

See also Jackson, Charles T., 1848b.

Foster, J. W., 1849a, Geological notes submitted to C. T. Jackson, SED 1, v. 3, 31-1, p. 766-771, (551).

Also in HED 5, v. 3, part 3, 31-1, p. 766-771, (571). Describes the geology of part of the Lake Superior region. See also Jackson, C. T., 1849.

Foster, J. W., 1849b, Notes on the geology and topography of portions of the country adjacent to Lake Superior and Michigan, SED 1, v. 3, 31-1, p. 773-785, (551).

Also in HED 5, v. 3, part 3, 31-1, p. 773-785, (571). See also Jackson, C. T., 1849.

Foster, J. W., and **Whitney, J. D.**, 1849, Synopsis of the explorations of the geological corps in the Lake Superior land district in the northern peninsula of Michigan . . . , SED 1, v. 3, 31-1, p. 605-611, (551).

Also in HED 5, v. 3, part 3, 31-1, p. 605-611, (571). See also Jackson, C. T., 1849.

Foster, J. W., and **Whitney, J. D.**, 1850a, Mineral report (on the Lake Superior region), SED 2, v. 2, 31-2, p. 147-152, (588).

Describes the field work completed during the past season.

Foster, J. W., and **Whitney, J. D.**, 1850b, Report on the geology and topography of a portion of the Lake Superior land district in the state of Michigan, part 1, Copper lands, HED 69, v. 9, 31-1, p. 3-224, 12 plates, 1 map, (578).

Describes in detail the geology, mines, and mining of the region. Part 2 is found in Foster, J. W., and Whitney, J. D., 1851.

Foster, J. W., and **Whitney, J. D.**, 1851, Report on the geology of the Lake Superior district, part 2, the Iron region, together with the general geology, SED 4, v. 3, 32-special, p. 1-406, 37 plates, 3 maps, (609).

Part 1 is found in Foster, J. W., and Whitney, J. D., 1850b.

Frazer, John F., 1850, (Report on minerals collected in Oregon and California), SED 47, part 1, v. 10, 31-1, p. 116-117, (558).

Describes the chemical analysis of a coal sample and a limestone sample collected during P. T. Tyson's 1849 geological exploration of California.

Fremont, John C., 1843, A report on an exploration of the country lying between the Missouri River and the Rocky Mountains . . . , SED 243, v. 4, 27-3, p. 1-207, 6 plates, 1 map, (416).

Includes descriptions of the landforms and lithology. Fremont, John C., 1845, includes this report and an additional report on an exploring expedition to Oregon and California.

Fremont, John C., 1845, Report of the exploring expedition to the Rocky Mountains in the year 1842, and to Oregon and California in the years 1843-44, SED 174, v. 11, 28-2, p. 1-693, 22 plates, 5 maps, (461).

Also in HED 166, v. 4, part 2, 28-2, p. 1-583, 22 plates, 5 maps, (467). See also Fremont, John C., 1843, and also Hall, James, 1845a, 1845b.

Fremont, John C., 1848, Geographical memoir upon Upper California in illustration of his map of Oregon and California, SMD 148, v. 1, 30-1, p. 1-67, 1 map, (511).

Also in HMD 5, v. 1, 30-2, p. 1-40, 1 map, (544).

Froebel, Julius, 1855, Remarks contributing to the physical geography of the North American continent, SMD 24, v. 1, 33-2, p. 272-281, (772).

Also in HMD 37, v. 1, 33-2, p. 272-281, (807). In the Annual Report of the Smithsonian Institution.

Fuertes, E. A., 1872, Report (on canal across the Isthmus of Tehuantepec), SED 6, v. 3, 42-2, p. 25-80, 11 plates, (1480).

Describes the engineering geology problems in canal construction in Vera Cruz and Oaxaca. See also Shufeldt, Robert W., 1872.

Fuller, Charles A., 1855, Survey and map of the Red River in the region of the raft, SED 62, v. 7, 33-2, p. 1-6, 1 map, (752).

Describes the natural log dam and the associated fluvial features.

Fuller, Myron L., Hydrologic work of the U. S. Geological Survey in the eastern United States, HD 460, v. 111, 58-3, p. 509-514, (4890).

In the Report of the 8th International Geographic Congress.

Gabb, W. M., 1867, Report on Coal, HED 29, v. 7, 39-2, p. 188-193, (1289).

Describes the known coal deposits west of the Rockies. See also Browne, J. R., 1867.

Gamble, Robert J., 1902, Wind Cave national park, SR 1944, v. 9, 57-1, p. 1-6, (4264).

Describes the caves in a proposed national park in South Dakota.

Gannett, E. M., 1883, Geographical field work of the Yellowstone Park division, HMD 62, part 2, v. 22, 47-1, p. 455-490, (2057).

Describes the topography of parts of Wyoming. See also Hayden, F. V., 1883b.

Gardner, Frank D., and **Jensen, Charles A.,** 1901a, Soil survey in the Sevier Valley, Utah, HD 526, v. 107, 56-2, p. 243-285, 4 maps, (4181).

Gardner, Frank D., and **Jensen, Charles A.,** 1901b, Soil survey in Weber County, Utah, HD 526, v. 107, 56-2, p. 207-242, 1 plate, 3 maps, (4181).

Gardner, Frank D., and **Stewart, John,** 1900, A soil survey in Salt Lake Valley, Utah, HD 399, v. 88, 56-1, p. 77-114, 11 plates, 4 maps, (3985).

Gaudry, Albert, 1903, The Baouseé-Roussé explorations: study of a new human type, by M. Verneau, HD 484, part 1, v. 109, 57-2, p. 451-453, 2 plates, (4548).

Describes hominid fossils from France. In the Annual Report of the Smithsonian Institution.

Gautier, E. F., 1905, La valeur commercial et industrielle du Sahara Francais, HD 460, v 111, 58-3, p. 892-900, (4890).

Describes the known mineral resources of Algeria. In the Annual Report of the Smithsonian Institution.

Geib, W. J., 1908, Soil survey of Cass County, Michigan, HD 352, v. 75, 59-2, p. 729-745, 1 map (in v. 76), (5178-5179).

Geib, W. J., and **Jones, Grove B.,** 1907, Soil survey of the Carlton area, Minnesota-Wisconsin, HD 925, v. 110, 59-1, p. 815-835, 1 map (in v. 111), (5050-5051).

Geikie, Archibald, 1882, Review of "The geology of the high plateaus of Utah" by C. E. Dutton, HED 1, part 2, v. 6, 47-1, p. 178-184, (2014).

Reviews a scientific publication resulting from explorations by the War Department. See also Dana, James D., 1882.

Geikie, Archibald, 1884, Review of "The Tertiary history of the Grand Canyon district" by Captain C. E. Dutton, HED 1, part 2, v. 6, 48-1, p. 128-132, (2185).

Reviews a scientific publication resulting from explorations by the War Department.

Geikie, Archibald, 1893, Geologic change and time, HMD 114, part 1, v. 22, 52-2, p. 111-131, (3131).

Discusses geochronology, uniformitarianism, and the age of the earth. In the Annual Report of the Smithsonian Institution.

Geikie, James, 1891, Glacial geology, HMD 129, part 1, v. 10, 51-2, p. 221-230, (2878).

Describes the general state of knowledge about Pleistocene geology. In the Annual Report of the Smithsonian Institution.

Geikie, James, 1899, The tundras and steppes of prehistoric Europe, HD 309, v. 91, p. 321-348, 1 map, (3833).

Describes the paleoclimatology of Quaternary Europe. In the Annual Report of the Smithsonian Institution.

Gerland, G., 1905, Gründung, Organisation, und Aufgaben der Internationalen Seismologischen Assoziation, HD 460, v. 111, 58-3, p. 468-477, (4890).

In the Report of the 8th International Geographic Congress.

Gibbon, Lardner, 1854, Exploration of the valley of the Amazon, part 2, HED 53, v. 9, 33-1, p. 1-339, 36 plates, 2 maps, (722).

Also in SED 36, v. 6, part 2, 33-2, p. 1-339, 36 plates, 2 maps, (664). For part one of the report see Herndon, Lewis, 1853. Describes the hydrology, topography, economic deposits, and mining of the area.

Gibbs, George, 1855, Report upon the geology of the central portion of Washington Territory, SED 78, v. 13, part 1, 33-2, p. 473-486, (758).

Also in HED 91, v. 11, part 1, 33-2, p. 473-486, (791); and in HED 129, v. 18, part 1, 33-1, p. 494-512, (736). Part of the Pacific Railroad Survey Reports. See also Stevens, I. I., 1855a.

Gibbs, William P., 1849, Field notes of . . ., SED 1, v. 3, 31-1, p. 702-711, (551).

Also in HED 5, v. 3, part 3, 31-1, p. 702-711, (571). Describes field work carried out in part of the Lake Superior region. See also Jackson, Charles T., 1849.

Gibbs, Oliver W., 1856, . . . results of examinations made of sands taken from . . . Key Biscayne, Cape Florida, and Cape Sable, SED 18, v. 15, 34-3, p. 318-319, (888).

Also in HED 18, v. 4, 34-3, p. 318-319, (898). In the Annual Report of the Coast Survey.

Gilbert, G. K., 1872, (Report of progress in connection with Wheeler's survey of Nevada and Arizona), SED 65, v. 2, 42-2, p. 90-94, (1479).

See also Wheeler, George M., 1872a.

Gilbert, G. K., 1885, The sufficiency of terrestrial rotation for the deflection of streams, SMD 69, v. 5, 48-2, p. 7-10, (2269).

In the Memoirs of the National Academy of Sciences, v. 3, part 1, Memoir 1.

Gilbert, G. K., 1891, The history of Niagra River, HMD 129, part 1, v. 10, 51-2, p. 231-257, 8 plates, (2878).

Dscribes the fluvial features and the geologic history of the river. In the Annual Report of the Smithsonian Institution.

Gilbert, G. K., 1893, Continental problems of geology, HMD 114, part 1, v. 22, 52-2, p. 163-173, (3131).

Discusses crustal instability, density, and structure, and the growth of continents, in the Annual Report of the Smithsonian Institution.

Gilbert, G. K., 1899, Modification of the Great Lakes by earth movements, HD 309, v. 91, 55-3, p. 349-361, (3833).

Describes the lacustrine and glacial features and their relation to isostatic rebound. In the Annual Report of the Smithsonian Institution.

Gilbert, G. K., 1903, John Wesley Powell, HD 484, part 1, v. 109, 57-2, p. 633-640, 1 plate, (4548).

Describes the life and work of Powell. In the Annual Report of the Smithsonian Institution.

Gilbert, G. K., 1905, The sculpture of massive rocks, HD 460, v. 111, 58-3, p. 191-192, (4890).

An abstract in the Report of the 8th International Geographic Congress.

Gill, Theodore N., 1896, Huxley and his work, HD 425, part 1, v. 81, 54-1, p. 759-779, (3448).

Describes the life and work of T. H. Huxley. In the Annual Report of the Smithsonian Institution.

Gill, Theodore N., 1905, Origin of freshwater faunas, HD 460, v. 111, 58-3, p. 617, (4890).

An abstract in the Report of the 8th International Geographic Congress.

Gilliss, J. M., 1855, Chile: its geography, climate, earthquakes, governments, social condition, mineral and agricultural resources, commerce, etc., etc., HED 121, v. 15, part 1, 33-1, p. 1-556, 11 plates, 3 maps, (728).

Describes in detail the topography, mineral waters, mineral resources, and earthquakes. This forms v. 1 of the U.S. Astronomical Expedition to the Southern Hemisphere. See also Conrad, T. A., 1855c; Smith, J. Lawrence, 1855; and Wyman, Jeffries, 1855.

Goiticoa, N. V., 1904, Venezuela: geographical sketch, natural resources . . . , HD 145, part 3, v. 65, 58-3, p. 1-608, 35 plates, 1 map, (4844).

Describes the topography and mineral resources of the country.

Goode, George B., 1901, The genesis of the United States National Museum, HD 575, part 3, v. 79, part 2, 55-2, p. 83-191, 27 plates, (3708).

Describes the origin and growth of the museum. In the Annual Report of the Smithsonian Institution.

Gore, James H., 1903, A bibliography of geodesy, SD 223, v. 20, 57-2, p. 427-787, (4435).

In the Annual Report of the Coast and Geodetic Survey.

Governor of New Mexico, 1905, Geology (of New Mexico), HD 5, v. 21, 59-1, p. 419-422, (4961).

Briefly describes the general geology of the state.

Graham, George, 1829, Inundated lands on the Mississippi, HED 99, v. 3, 20-2, p. 1-11, (186).

Describes the fluvial features of the Mississippi River in Louisiana and Mississippi.

Gray, A. B., 1846, Mineral lands on Lake Superior, HED 211, v. 7, 29-1, p. 2-23, 1 map, (486).

Describes the topography and economic minerals of part of the Lake Superior region.

Gregory, J. W., 1892, Final report on the mid-plains division of the artesian and underflow investigation . . . , SED 41, part 4, v. 4, 52-1, p. 1-50, 20 plates and maps, (2899).

Describes the areal geology and the ground water conditions of parts of Nebraska, Oklahoma, Kansas, and Colorado.

Gregory, J. W., 1899, The plan of the earth and its causes, HD 309, v. 91, 55-3, p. 363-388, (3833).

Discusses the shape and crustal features of the earth and the permanence of continents and ocean basins. In the Annual Report of the Smithsonian Institution.

Griffen, A. M., and **Ayrs, Orla L.,** 1907a, Soil survey of Madison County, Kentucky, HD 925, v. 110, 59-1, p. 659-678, 1 map (in v. 111), (5050-5051).

Griffen, A. M., and **Ayrs, Orla L.,** 1907b, Soil survey of Upshur County, West Virginia, HD 925, v. 110, 59-1, p. 175-190, 1 map (in v. 111), (5050-5051).

Griffen, A. M., and **Caine, Thomas A.,** 1907, Soil survey of Tangipahoa Parish, Louisiana, HD 925, v. 110, 59-1, p. 493-515, 1 map (in v. 111), (5050-5051).

Griffen, A. M., et al, 1908, Soil survey of Escambia County, Florida, HD 352, v. 75, 59-2, p. 335-362, 1 map (in v. 76), (5178-5179).

Grinell, George B., 1874, Preliminary report on paleontology . . . (of the Black Hills), HED 1, part 2, v. 4, 43-2, p. 632-633, (1637).

A brief description of the invertebrates. See also Winchell, N. H., 1874.

Grinell, George B., 1875, Paleontological report (on the Black Hills), HED 1, part 2, v. 5, 44-1, p. 1177-1180 and 1203, 1 plate, (1676).

Describes and illustrates the invertebrates. See also Ludlow, William, 1875.

Gulliver, F. P., 1905, Island typing, HD 460, v. 111, 58-3, p. 146-149, (4890).

Describes the coastal features and sea level changes in parts of Massachusetts. In the Report of the 8th International Geographic Congress.

Haeckel, Ernst, 1899, On our present knowledge of the origin of man, HD 309, v. 91, 55-3, p. 461-480, (3833).

In the Annual Report of the Smithsonian Institution.

Hague, Arnold, 1893a, Geological history of the Yellowstone National Park, HMD 114, part 1, v. 22, 52-2, p. 133-151, (3131).

Describes the geology and thermal spring activity in the park. In the Annual Report of the Smithsonian Institution.

Hague, Arnold, 1893b, Soaping geysers, HMD 114, part 1, v. 22, 52-2, p. 153-161, (3131).

Describes the geysers in Yellowstone National Park and the tourist practice of "soaping" to induce geyser activity. In the Annual Report of the Smithsonian Institution.

Hall, Charles F., 1879, Narrative of the 2nd Arctic expedition . . . during the years 1864-69, SED 27, v. 3, 45-3, p. 1-644, 8 plates, 13 maps, (1830).

Describes the ice features and landforms of parts of the Northwest and Arctic Territories. See also Emerson, Benjamin K., 1879.

Hall, James, 1845a, Descriptions of organic remains collected by Captain Fremont in his geographical survey of Oregon and North California, SED 174, v. 11, 28-2, p. 304-310, 4 plates, (461).

Also in HED 166, v. 4, part 2, 28-2, p. 304-310, 4 plates, (467). Describes the invertebrates and plants collected by Fremont.

Hall, James, 1845b, Nature of the geological formations occupying the portion of Oregon and North California included in a geographical survey under the direction of Captain Fremont, SED 174, v. 11, 28-2, p. 295-303, 1 plate, (461).

Also in HED 166, v. 4, part 2, 28-2, p. 295-303, 1 plate, (467). Describes the rock samples collected by Fremont.

Hall, James, 1851a, Description of new and rare species of fossils from the Paleozoic Series (of the Lake Superior region), SED 4, v. 3, 32-special, p. 203-231, 14 plates, (609).

List and descriptions of fossils that form chapter 13 of Foster, J. W., and Whitney, J. D., 1851.

Hall, James, 1851b, Lower Silurian, Upper Silurian, and Devonian Series (of the Lake Superior region), SED 4, v. 3, 32-special, p. 140-166, (609).

Describes the Silurian and Devonian stratigraphy of the area. Forms chapter 9 of Foster, J. W., and Whitney, J. D., 1851.

Hall, James, 1851c, Parallelism of the Paleozoic deposits of Europe and America, SED 4, v. 3, 32-special, p. 285-318, (609).

Correlates the sedimentary sequences in the two areas. Found in chapter 18 of Foster, J. W., and Whitney, J. D., 1851.

Hall, James, 1852, Observations on the geology and palentology of the country traversed by the (Stansbury) expedition, SED 3, v. 2, 32-special, p. 401-414, 4 plates, (608).

Describes invertebrates collected during an exploration of the west. See also Stansbury, Howard, 1852. Plates are not found in all copies of this Senate edition. Some copies of this Senate edition bear the date 1853.

Hall, James, 1855, Descriptions and notices of the fossils collected upon the route (near the 35th parallel), SED 78, v. 13, part 3, 33-2, p. 100-105, 2 plates, (760).

Also in HED 91, v. 11, part 3, 33-2, p. 100-105, 2 plates, (793). Part of the Pacific Railroad Survey Reports. Forms chapter 9 of part 4 of Whipple, A. W., 1855.

Hall, James, 1857, Geology and paleontology of the (United States and Mexican) boundary, SED 108, v. 20, part 1, 34-1, p. 101-174, 21 plates, (832).

Also in HED 135, v. 14, part 1, 34-1, p. 101-174, 21 plates, (861). Describes the rock specimens and fossils collected during the survey and correlates the rocks with other parts of the country. See also Parry, C. C., and Schott, Arthur, 1857.

Hallock, William, 1893, The flow of solids, or the behavior of solids under high pressure, HMD 334, part 1, v. 43, 52-1, p. 237-246, (3001).

Describes the laboratory deformation of rocks. In the Annual Report of the Smithsonian Institution.

Hamilton, Pierce S., 1868, Gold and coal mines of Nova Scotia, HED 273, v. 17, 40-2, p. 56-68, (1343).

See also Taylor, James W., 1868.

Hammond, Harry, 1884, Physio-geographical and agricultural features of the state of South Carolina, HMD 42, part 6, v. 13, part 6, 47-2, p. 5-49, 3 plates and maps, (2134).

> In the reports of the 10th Census. Describes the landforms and soils of the state.

Harshberger, J. W., 1905, Methods of determining the age of the different floristic elements of eastern North America, HD 460, v. 111, 58-3, p. 601-607, (4890).

> In the Report of the 8th International Geographic Congress.

Hatch, F. H., and **Corstorphine, G. S.,** 1906, The Cullinan diamond–a description of the big diamond found in the Premier mine, Transvaal, HD 930, v. 113, 59-1, p. 211-213, 2 plates, (5053).

> In the Annual Report of the Smithsonian Institution.

Hatfield, Chester, 1874, Report of explorations and surveys for the location of a ship canal . . . through Nicaragua, SED 57, v. 3, 43-1, p. 1-143, 20 plans and maps, (1582).

> Describes the exploration and topography of parts of the country. See also Endlich, Frederick, 1874; and Whitfield, Benjamin, 1874.

Hawes, George W., 1881a, Record of recent scientific progress in geology, SMD 31, v. 1, 46-3, p. 221-234, (1944).

> In the Annual Report of the Smithsonian Institution.

Hawes, George W., 1881b, Record of recent scientific progress in mineralogy, SMD 31, v. 1, 46-3, p. 299-312, (1944).

> In the Annual Report of the Smithsonian Institution.

Hawes, George W., 1884, Report on the building stones of the United States . . . , HMD 42, part 10, v. 13, part 10, 47-2, p. 1-399, 58 plates, (2139).

> A complete description that includes chemistry, microscopic structure, and quarries and quarry regions. A part of the reports of the 10th Census.

Hawn, F., 1874a, Geological description, examination of the Musca Pass from Fort Garland to Huerfano Valley, HED 193, v. 12, 43-1, p. 62-66, (1610).

> Describes the general geology of part of Colorado. In Ruffner, E. H., 1874.

Hawn, F., 1874b, Geological observations . . . from Fisch's Ranch to Pueblo . . . , HED 193, v. 12, 43-1, p. 75-85, (1610).

Describes the general geology of part of Colorado. In Ruffner, E. H., 1874.

Hawn, F., 1874c, Geological observations from the mouth of Lake Mary Canyon . . . to Del Norte, HED 193, v. 12, 43-1, p. 73-75, (1610).

Describes the general geology of part of Colorado. In Ruffner, E. H., 1874.

Hawn, F., 1874d, Report of the geological survey of the Los Animas mining district . . . , HED 193, v. 12, 43-1, p. 69-73, (1610).

Describes the mineral resources of this part of Colorado. In Ruffner, E. H., 1874.

Hawn, Laurens, 1874, Notes made on the Ute reconnaissance during examination of the Animas River . . . , HED 193, v. 12, 43-1, p. 66-69, (1610).

Describes the general geology of part of Colorado. In Ruffner, E. H., 1874.

Hay, Robert, 1892, Final geological reports on the artesian and underflow investigation between the 97th meridian . . . and the foothills of the Rocky Mountains, SED 41, part 3, v. 4, 52-1, p. 1-39, 18 plates and maps, (2899).

Describes the areal geology and the ground water conditions of the Great Plains.

Hayden, F. V., 1858, Catalogue of the collections in geology and natural history obtained in Nebraska and portions of Kansas, during several expeditions (under Lt. G. J. Warren), SED 1, v. 3, 35-2, p. 673-747, (976).

The catalogue includes rock and fossil specimens. The general stratigraphy and specific stratigraphic sections for parts of Kansas and Nebraska are also included.

Hayden, F. V., 1868, Lignite of the west, HED 273, v. 17, 40-2, p. 35-38, (1343).

Describes some recently discovered coal deposits. In Taylor, James W., 1868.

Hayden, F. V., 1871, Final report of the United States Geological Survey of Nebraska and portions of adjacent territories, HED 19, v. 2, 42-1, p. 1-264, 11 plates, (1471).

See also Meek, F. P., 1871; and St. John, Orestes, H., 1871.

Hayden, F. V., 1872a, Preliminary report of the United States geological survey of Montana and portions of adjacent territories, HED 326, v. 15, 42-2, p. 1-538, 2 plates, 5 maps, (1520).

Hayden's 5th annual progress report on this area. See also Cope, E. D., 1872c, 1872d; Leidy, Joseph, 1872a; Lesquereux, L., 1872a; Meek, F. B., 1872b; Peale, A. C., 1872; and Thomas, Cyrus, 1872.

Hayden, F. V., 1872b, Preliminary report of the United States geological survey of Wyoming and portions of contiguous territories, HED 325, v. 15, 42-2, p. 1-511, (1520).

Hayden's 2nd annual progress report on this area. See also Cope, E. D., 1872a, 1872b; Elliott, R. S., 1872; Hodge, Joseph T., 1872; Leidy, Joseph, 1872b; Lesquereux, L., 1872b; Meek, F. B., 1872c; and Newberry, J. S., 1872.

Hayden, F. V., 1873, Sixth annual report of the United States geological survey of the territories, embracing portions of Montana, Idaho, Wyoming, and Utah . . . , HMD 112, v. 3, 42-3, p. 1-844, 12 plates, 5 maps, (1573).

A progress report of the work under Hayden's direction. See also Bannister, H. M., 1873; Bradley, F. H., 1873; Cope, E. D., 1873; Langford, N. P., 1873a; Lesquereux, L., 1873; Meek, F. B., 1873; Peale, A. C., 1873; and Thomas, Cyrus, 1873.

Hayden, F. V., 1877, Preliminary report of the field work of the United States geological and geographical survey of the territories, HED 1, part 5, v. 8, 45-2, p. 755-787, (1800).

A progress report that describes the geology of parts of Colorado, Idaho, Utah, and Wyoming.

Hayden, F. V., 1878a, (Letter regarding geological and geographical surveys), HED 81, v. 17, 45-2, p. 1-19, 1 map, (1809).

A letter to the Secretary of the Interior describing Hayden's work of the preceding 10 years. See also Powell, John W., 1878a; and Humphreys, A. A., 1878b.

Hayden, F. V., 1878b, Preliminary report of the field work of the United States geological and geographical survey of the territories for the season of 1878, HED 1, part 5, v. 9, 45-3, p. 951-956, (1850).

A report of the progress of the work under Hayden's direction.

Hayden, F. V., 1878c, The primordial sandstone of the Rocky Mountains in the northwestern territories of the United States, HED 1, part 5, v. 9, 45-3, p. 964-969, (1850).

A reprint of an article from the American Journal of Science for January 1862.

Hayden, F. V., 1878d, Some remarks in regard to the period of elevation of those ranges of the Rocky Mountains near the source of the Missouri River . . . , HED 1, part 5, v. 9, 45-3, p. 957-961, (1850).

A reprint of an article from the American Journal of Science for May 1862.

Hayden, F. V., 1878e, Remarks on the geological formations along the eastern margins of the Rocky Mountains, HED 1, part 5, v. 9, 45-3, p. 961-964, (1850).

A reprint of an article from the American Journal of Science for May 1868.

Hayden, F. V., 1878f, Sketch of the geology of the country about the headwaters of the Missouri and Yellowstone Rivers, HED 1, part 5, v. 9, 45-3, p. 969-977, (1850).

A reprint of an article from the American Journal of Science for March 1861.

Hayden, F. V., 1883a, Twelfth annual report of the United States geological and geographical survey of the territories . . . , part 1, HMD 62, part 1, v. 21, 47-1, p. 1-786, 154 plates, 7 maps, (2056).

Describes the exploration and invertebrate paleontology of parts of Wyoming, Idaho, Utah, and Colorado, and the invertebrate paleontology of parts of Iowa, Illinois, Missouri, and Indiana. See also Hayden, F. V., 1883b.

Hayden, F. V., 1883b, Twelfth annual report of the United States geological and geographical survey of the territories . . . , part 2, HMD 62, part 2, v. 22, 47-1, p. 1-503, (2057).

Describes the exploration and geology of parts of Wyoming, especially Yellowstone National Park. See also Hayden, F. V., 1883a.

Hayes, C. Willard, 1906, Physiography of the Isle of Pines, SD 311, v. 6, 59-1, p. 35-37, (4914).

Describes the topography of this West Indian island.

Hays, S. S., 1866, Report . . . on petroleum, HED 51, v. 8, 39-1, p. 1-39, (1256).

Describes the petroleum resources of the U. S. and the rest of the world, and describes the history of petroleum in the U. S.

Haywood, J. K., 1902, The chemical composition of the waters of the Hot Springs of Arkansas, with an account of the methods of analysis . . . , SD 282, v. 20, 57-1, p. 4-78, (4239).

See also Weed, Walter H., 1902.

Hearn, W. Edward, 1903, Soil survey of the Lyons area, New York, HD 473, v. 106, 57-2, p. 143-162, 2 plates, 1 map, (4545).

Hearn, W. Edward, 1904, Soil survey of the Stanton area, Nebraska, HD 746, v. 111, 58-2, p. 947-962, 1 map (in v. 112), (4737-4738).

Hearn, W. Edward, and **Burgess, James L.,** 1904a, Soil survey of the Grand Island area, Nebraska, HD 746, v. 111, 58-2, p. 927-945, 1 map (in v. 112), (4737-4738).

Hearn, W. Edward, and **Burgess, James L.,** 1904b, Soil survey of the Jacksonville area, Texas, HD 746, v. 111, 58-2, p. 521-531, 1 map (in v. 112), (4737-4738).

Hearn, W. Edward, and **Burgess, James L.,** 1904c, Soil survey of the Nacogdoches area, Texas, HD 746, v. 111, 58-2, p. 487-499, 1 map (in v. 112), (4737-4738).

Hearn, W. Edward, and **Carr, M. E.,** 1905, Soil survey of the Biloxi area, Mississippi, HD 458, v. 108, 58-3, p. 353-374, 1 map (in v. 109), (4887-4888).

Hearn, W. Edward, and **Geib, W. J.,** 1908, Soil survey of Lee County, Alabama, HD 352, v. 75, 59-2, p. 363-384, 1 map (in v. 76), (5178-5179).

Hearn, W. Edward, and **Griffen, A. M.,** 1905, Soil survey of the Alma area, Michigan, HD 458, v. 108, 58-3, p. 639-664, 1 map (in v. 109), (4887-4888).

Hearn, W. Edward, and **MacNider, G. M.,** 1908a, Soil survey of Chowan County, North Carolina, HD 352, v. 75, 59-2, p. 223-244, 1 map (in v. 76), (5178-5179).

Hearn, W. Edward, and **MacNider, G. M.,** 1908b, Soil survey of Transylvania County, North Carolina, HD 352, v. 75, 59-2, p. 281-301, 1 map (in v. 76), (5178-5179).

Hearn, W. Edward, and **Mann, Charles J.,** 1907a, Soil survey of Crawford County, Missouri, HD 925, v. 110, 59-1, p. 865-878, 1 map (in v. 111), (5050-5051).

Hearn, W. Edward, and **Mann, Charles J.,** 1907b, Soil survey of Scotland County, Missouri, HD 925, v. 110, 59-1, p. 879-892, 1 map (in v. 111), (5050-5051).

Hearn, W. Edward, et al, 1904, Soil survey of the Lufkin area, Texas, HD 746, v. 111, 58-2, p. 501-510, 1 map (in v. 112), (4737-4738).

Heileman, W. H., and **Mesmer, Louis,** 1902, Soil survey of the Lake Charles area, Louisiana, HD 655, v. 117, 57-1, p. 621-647, 1 map, (4384).

Heilprin, Angelo, 1905a, Destruction of Pompeii as interpreted by the volcanic eruptions of Martinique, HD 460, v. 111, 58-3, p. 445, (4890).

An abstract in the Report of the International Geographic Congress.

Heilprin, Angelo, 1905b, Tower of Pelee, HD 460, v. 111, 58-3, p. 446, (4890).

An abstract in the Report of the 8th International Geographic Congress.

Heizmann, C. L., 1874, Report on mineral and thermal waters, HED 285, v. 17, 43-1, p. 185-199, (1615).

Describes the thermal springs of Wyoming. In Jones, William A., 1874.

Hennessey, Henry, 1891, On the physical structure of the earth, HMD 129, part 1, v. 10, 51-2, p. 201-219, (2878).

Describes the internal structure of the earth and the nature of the core. In the Annual Report of the Smithsonian Institution.

Henry, Joseph, 1856, Circular relative to earthquakes, SMD 73, v. 1, 34-1, p. 245, (835).

A brief form giving instructions on what to report when describing earthquakes. In the Annual Report of the Smithsonian Institution.

Henry, Joseph, 1871, Eulogy on Prof. Alexander Dallas Bache, HED 20, v. 2, 42-1, p. 91-116, (1471).

Describes the life and work of the former head of the Coast Survey. In the Annual Report of the Smithsonian Institution.

Hergesheimer, E., 1883, Type forms of topography, Columbia River, SED 49, v. 3, 47-1, p. 124-125, (1988).

Describes the landforms along the Columbia Valley. In the Annual Report of the Coast and Geodetic Survey.

Herndon, Lewis, 1853, Exploration of the valley of the Amazon, part 1, SED 36, v. 6, part 1, 32-2, p. 5-414, 16 plates, 3 maps, (663).

Also in HED 43, v. 5, 32-2, p. 5-414, 16 plates, 3 maps (678). Describes the fluvial and topographic features and the economic geology of this part of Brazil. For part 2 see Gibbon, Lardner, 1854.

Herrick, C. L., 1900, Geological associations in New Mexico mining camps, HD 5, v. 30, 56-2, p. 387-390, (4104).

Describes the general geology and mineral deposits of parts of the state.

Herrick, C. L., 1901, Geology of the New Mexico oil fields, HD 5, v. 27, 57-1, p. 389-393, 1 plate, (4294).

Describes the general geology and oil fields of parts of New Mexico.

Herz, O. F., 1904, Frozen mammoth in Siberia, HD 748, v. 114, 58-2, p. 519-535, (4740).

Describes the finding and excavation of the mammoth. In the Annual Report of the Smithsonian Institution.

Hicks, L. E., 1892, On the underflow and sheet waters, irrigable lands, and geologic structure of Nebraska, SED 41, part 3, v. 4, 52-1, p. 167-190, 6 plates and maps, (2899).

Describes the areal geology and ground water conditions of Nebraska.

Hilgard, Eugene W., 1884a, General discussion on the cotton production of the United States . . . , HMD 42, part 5, v. 13, part 5, 47-2, p. 1-81, 4 plates and maps, (2133).

In the reports of the 10th Census. Describes the landforms of the Mississippi Valley and the soils of the cotton states.

Hilgard, Eugene W., 1884b, Physio-geographical and agricultural features of the state of California, HMD 42, part 6, v. 13, part 6, 47-2, p. 5-85, 1 map, (2134).

In the reports of the 10th Census. Describes the landforms and the soils of the state.

Hilgard, Eugene W., 1884c, Physio-geographical and agricultural features of the state of Louisiana, HMD 42, part 5, v. 13, part 5, 47-2. p. 9-32, 2 maps, (2133).

In the reports of the 10th Census. Describes the landforms and the soils of the state.

Hilgard, Eugene W., 1884d, Physio-geographical and agricultural features of the state of Mississippi, HMD 42, part 5, v. 13, part 5, 47-2, p. 7-87, 2 maps, (2133).

In the reports of the 10th Census. Describes the landforms and soils of the state.

Hilgard, Eugene W., and **Hopkins, F. V.**, 1878, (Report on the specimens collected from the borings made in 1874, between the Mississippi River and Lake Borgne), HED 1, part 2, v. 4, 45-3, p. 855-891, 4 plates, 1 map, (1845).

Describes the deltaic and fluvial features of the area, and the sediments, foraminifera and diatoms from the borings.

Hilgard, J. E., 1872, Abstract of a paper . . . on the earthquake-wave of August 14, 1868, HED 206, v. 8, 41-2, p. 233-234, (1419).

Describes coastal changes caused by tidal currents. In the Annual Report of the the Coast Survey.

Hilgard, J. E., 1875, On tides and tidal action in harbors, HMD 56, v. 2, 43-2, p. 207-226, (1654).

Describes coastal changes caused by tidal currents. In the Annual Report of the Smithsonian Institution.

Hilgard, J. E., 1885, Description of a model of the depths of the sea in the bay of North America and the Gulf of Mexico, HED 43, v. 22, 48-2, p. 619-621, 1 map, (2297).

Describes the known features of the continental shelf. In the Annual Report of the Coast and Geodetic Survey.

Hill, Robert T., 1892, On the occurrence of artesian and other underground waters in Texas, eastern New Mexico, and the Indian Territory west of the 97th meridian, SED 41, part 3, v. 4, 52-1, p. 41-166, 19 plates and maps, (2899).

Describes the areal geology and the ground water conditions for parts of Texas and New Mexico.

Hill, Robert T., 1905a, The physical geography of Mexico, HD 460, v. 111, 58-3, p. 765-766, (4890).

An abstract in the Report of the 8th International Geographic Congress.

Hill, Robert T., 1905b, Physical history of the Windward Islands, as illustrated in the larger story of Pelée—a study of volcanic and oceanic geography, HD 460, v. 111, 58-3, p. 244-245, (4890).

An abstract in the Report of the 8th International Geographic Congress.

Hinton, Richard J., 1890, A report on the preliminary investigation to determine the proper location of artesian wells within the area of the 97th meridian and east of the foothills of the Rocky Mountains, SED 222, v. 12, 51-1, p. 1-389, 33 plates and maps, (2689).

Describes the areal geology and ground water conditions for parts of India, Australia, Algeria, the Great Plains, and the Rocky Mountains.

Hinton, Richard J., 1892, A report on irrigation and cultivation of the soil . . . , SED 41, part 1, v. 4, 52-1, p. 1-457, 41 plates and maps, (2899).

Describes the ground water resources of parts of the Great Plains, Great Basin, Rocky Mountains, and the Pacific Coast.

Hitchock, C. H., 1874, The coal measures of the United States, HMD–unnumbered, v. 3, 42-1, p. 12-14, 2 maps, (1473).

In the reports of the 9th Census.

Hitchock, C. H., and **Blake, W. P.,** 1874, The geological map of the United States, HMD–unnumbered, v. 3, 42-1, p. 6-9, 2 maps, (1473).

In the reports of the 9th Census.

Hitchock, Edward, 1853, Notes upon the specimens of rocks and minerals collected (during an exploration of the Red River), SED 54, v. 8, 32-2, p. 163-178, (666).

Describes the lithology and geology on the basis of specimens collected during the exploration.

Hobbs, William H., 1902, Emigrant diamonds in America, HD 707, part 1, v. 126, 57-1, p. 359-366, 2 plates, 1 map, (4393).

Describes the origin of the diamond deposits in the United States. In the Annual Report of the Smithsonian Institution.

Hobbs, William H., 1905, Lineaments of the Atlantic border region, HD 460, v. 111, 58-3, p. 193-203, 1 map, (4890).

Describes the structural geology and the fault lineaments of the Continental shelf. In the Report of the 8th International Geographic Congress.

Hodge, Joseph T., 1872, On the Tertiary coals of the West, HED 325, v. 15, 42-2, p. 318-329, (1520).

A part of Hayden, F. V., 1872b.

Hoffman, William, 1879, Report of the reconnaissance of the Moreau or Owl River, Dakota, in 1878, HED 1, part 2, v. 5, 46-2, p. 2367-2369, (1906).

Describes the exploration and landforms of parts of South Dakota.

Holdich, T. H., 1903, The progress of geographical knowledge, HD 484, part 1, v. 109, 57-2, p. 351-373, (4548).

Includes a description of South American glaciation. In the Annual Report of the Smithsonian Institution.

Hollister, George B., 1905, Hydrographic work of the U. S. Geological Survey, HD 460, v. 111, 58-3, p. 515-522, (4890).

In the Report of the 8th International Geographic Congress.

Holmes, J. Garnett, 1901, Soil survey around Santa Ana, Cal., HD 526, v. 107, 56-2, p. 385-412, 8 plates, 2 maps, (4181).

Holmes, J. Garnett, 1902, Soil survey of the San Gabriel area, California, HD 655, v. 117, 57-1, p. 559-586, 1 map, (4384).

Holmes, J. Garnett, 1903a, Soil survey of the Walla Walla area, Washington, HD 473, v. 106, 57-2, p. 711-728, 2 maps, (4545).

Holmes, J. Garnett, 1903b, Soil survey of the Yuma area, Arizona, HD 473, v. 106, 57-2, p. 777-791, 3 plates, 2 maps, (4545).

Holmes, J. Garnett, 1904, Soil survey of the San Luis Valley, Colorado, HD 746, v. 111, 58-2, p. 1099-1119, 2 maps (in v. 112), (4737-4738).

Holmes, J. Garnett, and Mesmer, Louis, 1902, Soil survey of the Ventura area, California, HD 655, v. 117, 57-1, p. 521-557, 7 plates, 2 maps, (4384).

Holmes, J. Garnett, and Neill, N. P., 1905, Soil survey of the Greeley area, Colorado, HD 458, v. 108, 58-3, p. 951-993, 1 map (in v. 109), (4887-4888).

Holmes, J. Garnett, and **Rice, Thomas D.,** 1907, Soil survey of the Grand Junction area, Colorado, HD 925, v. 110, 59-1, p. 949-974, 2 maps (in v. 111), (5050-5051).

Holmes, J. Garnett, et al, 1904a, Soil survey of the Imperial area, California, HD 746, v. 111, 58-2, p. 1219-1248, 2 maps (in v. 112), (4737-4738).

Holmes, J. Garnett, et al, 1904b, Soil survey of the Indio area, California, HD 746, v. 111, 58-2, p. 1249-1262, 2 maps (in v. 112), (4737-4738).

Holmes, J. Garnett, et al, 1905a, Soil survey of the San Bernadino Valley, California, HD 458, v. 108, 58-3, p. 1115-1151, 1 map (in v. 109), (4887-4888).

Holmes, J. Garnett, et al., 1905b, Soil survey of the Yuma area, Arizona-California, HD 458, v. 108, 58-3, p. 1025-1047, 2 maps (in v. 109), (4887-4888).

Holmes, William H., 1883, Report on the geology of Yellowstone National Park, HMD 62, part 2, v. 22, 47-1, p. 1-62, 33 plates, (2057).

In Hayden, F. V., 1883b.

Holmes, William H., 1901, Review of the evidence relating to auriferous gravel man in California, HD 737, part 1, v. 119, 56-1, p. 419-472, 16 plates, (4016).

In the Annual Report of the Smithsonian Institution.

Holmes, William H., 1903a, Flint implements and fossil remains from a sulphur spring at Afton, Indian Territory, HD 707, part 2, v. 127, 57-1, p. 233-252, 26 plates, (4394).

Describes implements and hominid remains from Oklahoma. In the Annual Report of the U. S. National Museum.

Holmes, William H., 1903b, Fossil human remains found near Lansing, Kansas, HD 484, part 1, v. 109, 57-2, p. 455-462, 3 plates, (4548).

In the Annual Report of the Smithsonian Institution.

Houghton, Douglas, 1832, Report . . . on the copper of Lake Superior, HED 152, v. 4, 22-1, p. 17-20, (219).

Describes the native copper and other copper ores collected in 1831.

House Committee on Public Lands, 1874, Report on the geographical and geological surveys west of the Mississippi, HR 612, v. 4, 43-1, p. 1-91, (1626).

Letters and testimony dealing with the early western surveys. Information presented here influenced Congress to establish the U. S. Geological Survey.

Hovey, Edmund O., 1905a, Volcanoes of Martinique, Guadeloupe, and Saba, HD 460, v. 111, 58-3, p. 447-451, (4890).

Describes some of the West Indian volcanoes. In the Report of the 8th International Geographic Congress.

Hovey, Edmund O., 1905b, Volcanoes of St. Vincent, St. Kitts, and Statia, HD 460, v. 111, 58-3, p. 452-454, (4890).

Describes some of the West Indian volcanoes. In the Report of the 8th International Geographic Congress.

Howard, C. W., 1867, Condition and resources of Georgia, HED 107, v. 15, 39-2, p. 567-580, (1297).

Describes the general geology and the mineral resources of the state. In the Report of the Commissioner of Agriculture.

Howard, William, 1828, Report (on) a route for a national road from the city of Washington to the northwestern frontier of the state of New York, HED 38, v. 2, 20-2, p. 2-22, 2 maps, (185).

Describes the landforms and the lithology of the proposed route.

Hubbard, Bela, 1846, General observations on the geology and topography of the district south of Lake Superior . . . , SED 357, v. 7, 29-1, p. 20-29, (476).

Includes a stratigraphic section for Upper Michigan.

Hughes, George W., 1837, A report of the survey of the harbor of Havre-de-Grace, HED 134, v. 3, 24-2, p. 1-10, (303).

Describes the coastal features and coastal changes at the head of Chesapeake Bay. Sediments here are compared with those in other parts of the world.

Hughes, George W., 1844, Report relative to the working of Copper ore, SED 291, v. 5, 28-1, p. 1-58, (435).

Describes copper mining and ore treatment as practiced in Cornwall and Devon, England.

Hughes, George W., et al, 1837, Petition . . . praying Congress to establish a Mineralogical Cabinet . . . , SED 167, v. 2, 24-2, p. 1-2, (298).

Humphreys, A. A., 1878a, Letter . . . giving information concerning the geographical and geological surveys of the War Department, HED 1, part 2, v. 5, 45-3, p. 1653-1660, (1846).

Describes the history of these surveys. See also United States War Department, 1878.

Humphreys, A. A., 1878b, Surveys by the War Department, HED 88, v. 17, 45-2, p. 1-8, 1 map, (1809).

Describes the surveys of the previous 10 years. See also Powell, John W., 1878a; and Hayden, F. V., 1878a.

Humphreys, A. A., and **Warren, G. K.,** 1855, An examination . . . of the reports of explorations for railroad routes from the Mississippi to the Pacific . . . , in 1853-'54 . . . , SED 78, v. 13, part 1, 33-2, p. 35-111, (758).

Also in HED 91, v. 11, part 1, 33-2, p. 35-111, (791); *and* HED 129, v. 18, part 1, 33-1, p. 1-96, (736). Part of the Pacific Railroad Survey Reports. Describes the general features and construction costs of each of the routes.

Hunt, E. B., 1864, On the origin, growth, substructure, and chronology of the Florida reef, HED 22, v. 9, 37-3, p. 241-248, (1165).

Describes the reef features and dates the reef by the rate of coral growth. In the Annual Report of the Coast Survey.

Hunt, T. Sterry, 1862, Notes on the history of petroleum or rock oil, SMD 77, v. 1, 37-2, p. 319-329, (1141).

Describes the stratigraphic and geographic distribution of oil. In the Annual Report of the Smithsonian Institution.

Hunt, T. Sterry, 1871, On the chemistry of the earth, HED 153, v. 12, 41-3, p. 183-207, (1460).

Summarizes Hunt's ideas as put foreward over the previous 12 years. In the Annual Report of the Smithsonian Institution.

Hunt, T. Sterry, 1884, Record of recent scientific progress in geology, HMD 26, v. 5, 47-2, p. 325-345, (2121).

In the Annual Report of the Smithsonian Institution.

Hunt, T. Sterry, 1885, Record of scientific progress in geology for 1883, HMD 69, v. 34, 48-1, p. 443-464, (2246).

In the Annual Report of the Smithsonian Institution.

Huxley, Thomas H., 1871, Principles and methods in Paleontology, HED 153, v. 12, 41-3, p. 363-388, (1460).

In the Annual Report of the Smithsonian Institution.

Ickis, E. M., and **Field, E. M.,** 1906, Report upon building stone near Manila, HD 2, v. 12, 59-1, p. 321-328, (4952).

Describes the building stone resources in this part of the Philippines.

Ives, Joseph C., 1858, Preliminary report of the Colorado exploring expedition, SED 1, v. 2, 35-2, p. 608-619, (975).

Describes the progress of the exploration, the general geology of the area, and the mineral resources.

Ives, Joseph C., 1861, Report upon the Colorado River of the west, explored in 1857 and 1858, HED 90, v. 14, 36-1, p. 1-131, 27 plates, 1 map, (1058).

Describes the fluvial, topographic, and lithologic features of the lower Colorado River. See also Newberry, J. S., 1861.

Jackson, Charles T., 1848a, Synopsis of a report of the geological survey of the mineral lands of the United States in Michigan, SED 2, v. 2, 30-1, p. 175-230, (504).

See also Channing, William F., 1848a, 1848b, 1848c; Locke, John, 1848; and Whitney, J. D., 1848a, 1848b. Printing of this was authorized in late 1857 and some bibliographies use this date.

Jackson, Charles T., 1848b, Synopsis of a report on the progress of the geological survey of the mineral lands of the United States in Michigan, SED 2, v. 2, 30-2, p. 185-191, (530).

A progress report on the geological exploration of this area.

Jackson, Charles T., 1849, Report on the geological and mineralogical survey of the mineral lands of the United States . . . , SED 1, v. 3, 31-1, p. 371-801, 26 plates, 6 maps, (551).

Also in HED 5, v. 3, part 3, 31-1, p. 371-801, 26 plates, 6 maps, (571). The final report on the geological exploration of the mineral lands in the Lake Superior region.

Jamison, J., 1853, Report on the qualities of the coal fields and coal mines on the western waters, SED 46, v. 7, 32-2, p. 1-6, (665).

Describes the known coal deposits in Pennsylvania, West Virginia, Ohio, Illinois, Missouri, and Kentucky.

Jenny, Walter P., 1875, Report of geological survey of the Black Hills, HED 1, part 5, v. 9, 44-1, p. 683-685, (1680).

Describes the general geology of part of the Black Hills.

Jenny, Walter P., 1876, Report on the mineral wealth, climate, and rainfall, and natural resources of the Black Hills of Dakota, SED 51, v. 1, 44-1, p. 3-71, 1 map, (1664).

Describes the exploration and general geology of the area.

Jensen, Charles A., 1902, Soil survey of the Yakima area, Washington, HD, 655, v. 117, 57-1, p. 389-419, 7 plates, 6 maps, (4384).

Jensen, Charles A., 1904, Soil survey of the Salem area, Oregon, HD 746, v. 111, 58-2, p. 1171-1182, 1 map (in v. 112), (4737-4738).

Jensen, Charles A., and **Mackie, W. W.,** 1904, Soil survey of the Baker City area, Oregon, HD 746, v. 111, 58-2, p. 1151-1170, 4 maps (in v. 112), (4737-4738).

Jensen, Charles A., and **Neill, N. P.,** 1903a, Soil survey of the Grand Forks area, North Dakota, HD 473, v. 106, 57-2, p. 643-663, 2 plates, 3 maps, (4545).

Jensen, Charles A., and **Neill, N. P.,** 1903b, Soil survey of the Billings area, Montana, HD 473, v. 106, 57-2, p. 665-687, 2 plates, 4 maps, (4545).

Jensen, Charles A., and **Olshausen, B. A.,** 1902, Soil survey of the Boise area, Idaho, HD 655, v. 117, 57-1, p. 421-446, 5 plates, 5 maps, (4384).

Jensen, Charles A., and **Strahorn, A. T.,** 1905, Soil survey of the Bear River area, Utah, HD 458, v. 108, 58-3, p. 995-1023, 3 maps (in v. 109), (4887-4888).

Johnson, Lawrence C., 1888, Report on the iron regions of northern Louisiana and eastern Texas, HED 195, v. 26, 50-1, p. 1-54, 1 map, (2558).

Describes the general geology and the iron ore deposits of the area.

Johnston, A. R., 1858, Journal of Captain A. R. Johnston, First Dragoons, HED 41, v. 4, 30-1, p. 565-614, (517).

Describes the landforms and some of the lithology of parts of New Mexico and California.

Johnston, J. E., et al, 1850, Reconnaissance of routes from San Antonio to El Paso, SED 64, v. 14, 31-1, p. 3-54, 1 map, (562).

Includes descriptions of the landforms and rocks.

Joly, J., 1901, An estimate of the geologic age of the earth, HD 737, part 1, v. 119, 56-1, p. 247-288, (4016).

Describes the method of using dissolved salts in the oceans as a means for determining the age of the earth. In the Annual Report of the Smithsonian Institution.

Jones, George, 1856a, Description of a mineral spring near Hakodadi, HED 97, v. 12, part 2, 33-2, p. 97-98, (803).

Also in SED 79, v. 14, part 2, 33-2, p. 97-98, (770). In v. 2 of the Report of the Perry expedition to Japan and China. See also Perry, M. C., 1856.

Jones, George, 1856b, Report of a geological exploration of Lew Chew, HED 97, v. 12, part 2, 33-2, p. 53-56, (803).

Also in SED 79, v. 14, part 2, 33-2, p. 53-56, (770). In v. 2 of the Report of the Perry expedition to Japan and China. See also Perry, M. C., 1856.

Jones, George, 1856c, Report on the coal regions of the island of Formosa, HED 97, v. 12, part 2, 33-2, p. 153-163, (803).

Also in SED 79, v. 14, part 2, 33-2, p. 153-163, (770). In v. 2 of the Report of the Perry expedition to Japan and China. See also Perry, M. C., 1856.

Jones, Grove B., and **Ayrs, Orla L.,** 1908, Soil survey of Racine County, Wisconsin, HD 352, v. 75, p. 791-811, 1 map (in v. 76), (5178-5179).

Jones, Grove B., and **Carr, M. E.,** 1904, Soil survey of the Fort Payne area, Alabama, HD 746, v. 111, 58-2, p. 355-371, 1 map (in v. 112), (4737-4738).

Jones, Grove B., and **Carr, M. E.,** 1907, Soil survey of the Oxford area, Michigan, HD 925, v. 110, 59-1, p. 731-745, 1 map (in v. 111), (5050-5051).

Jones, Grove R., and **Ruhlen, La Mott,** 1905, Soil survey of De Soto Parish, Louisiana, HD 458, v. 108, 58-3, p. 375-395, 1 map (in v. 109), (4887-4888).

Jones, James O., 1903, Recent seismic disturbances in Guatemala, etc., SD 131, v. 9, 57-2, p. 1-27, (4424).

Describes earthquakes and volcanic eruptions in Guatemala, Costa Rica, and Nicaragua. See also Chamberlain, P. W., 1903.

Jones, William A., 1872, Report . . . of a survey and explorations in the Uintah Mountains, Utah, HED 1, part 2, v. 8, 42-3, p. 1108-1118, (1559).

Describes the topography of the area and briefly mentions the known economic deposits.

Jones, William A., 1874, Report upon the reconnaissance of northwestern Wyoming made in the summer of 1873, HED 285, v. 17, 43-1, p. 1-210, 4 plates, 50 maps, (1615).

Describes the exploration and general geology of the area. See also Comstock, Theodore B., 1874; and Heizmann, C. L., 1874.

Judd, John W., 1893, The rejuvenescence of crystals, HMD 114, part 1, v. 22, 52-2, p. 281-288, (3131).

Describes the growth and development of crystals. In the Annual Report of the Smithsonian Institution.

Keith, Arthur, 1902, Topography and geology of the Southern Appalachians, SD 84, v. 11, 57-1, p. 111-122, 10 plates, (4229).

Describes the topography and geology of parts of North and South Carolina, Georgia, Tennessee, and Virginia.

Kelvin, William T., 1898, The age of the earth as an abode fitted for life, HD 575, part 1, v. 78, 55-2, p. 337-357, (3706).

Discusses geochronology from the standpoint of heat loss by the earth. In the Annual Report of the Smithsonian Institution.

Kemp, James, 1907, The problem of the metalliferous veins, HD 575, v. 97, 59-2, p. 187-206, (5200).

Discusses the origin of ore deposits. In the Annual Report of the Smithsonian Institution.

Kerr, W. C., 1884a, Physio-geographical and agricultural features of the state of North Carolina. HMD 42, part 6, v. 13, part 6, 47-2, p. 7-27, 3 plates and maps, (2134).

In the reports of the 10th Census. Describes the landforms and the soils of the state.

Kerr, W. C., 1884b, Physio-geographical and agricultural features of the state of Virginia, HMD 42, part 6, v. 13, part 6, 47-2, p. 5-11, (2134).

In the reports of the 10th Census. Describes the landforms and the soils of the state.

Keyes, Charles R., 1903, Geological formations of New Mexico, HD 5, v. 23, 58-2, p. 337-341, (4649).

Describes the stratigraphy and areal geology of the state.

Keyes, W. S., 1868, Mineral resources of the territory of Montana, HED 273, v. 17, 40-2, p. 38-56, (1343).

Describes the mines and economic geology of the state. In Taylor, James W., 1868.

King, Clarence, 1871, (Brief account of work on the geological exploration of the 40th parallel), HED 1, part 2, v. 2, 42-2, p. 1027-1030, (1504).

A progress report of explorations under King's command.

King, Clarence, 1873, Report . . . of the geological exploration of the 40th parallel, from Sierra Nevada to the eastern slope of the Rocky Mountains, HED 1, part 2, v. 2, 43-1, p. 1203-1210, (1598).

A progress report of explorations under King's command.

King, Clarence, 1874, Annual report . . . for the fiscal year ending June 30, 1874, HED 1, part 2, v. 4, 43-2, p. 477-480, (1637).

A progress report of explorations under King's command.

King, Clarence, 1875, Annual report for the fiscal year ending June 30, 1875, HED 1, part 2, v. 5, 44-1, p. 919-920, (1676).

A progress report of explorations under King's command.

King, Clarence, 1876, Annual report . . . for the fiscal year ending June 30, 1876, HED 1, part 2, v. 5, 44-2, p. 217-218, (1745).

A progress report of explorations under King's command.

King, Clarence, 1877, Annual report . . . for the fiscal year ending June 30, 1877, HED 1, part 2, v. 4, 45-2, p. 1207, (1796).

A progress report of explorations under King's command.

King, Clarence, 1878, Annual report . . . for the fiscal year ending June 30, 1878, HED 1, part 2, v. 5, 45-3, p. 1419, (1846).

A progress report of explorations under King's command.

King, Clarence, 1880, Letter (in response to an inquiry as to the duties and value of the United States Geological Survey), SMD 48, v. 1, 46-2, p. 1-3, (1890).

Gives King's reasons for urging passage of a resolution which would not limit the U. S. G. S. to the "Public Domain".

King, Clarence, 1894, The age of the earth, HMD 184, part 1, v. 29, 53-2, p. 335-352, (3257).

Discusses the measurement of geologic time by the rate of cooling of the earth. In the Annual Report of the Smithsonian Institution.

King, Clarence, Emmons, S. F., and **Becker, G. F.,** 1885, Statistics and technology of the precious metals, HMD 42, part 13, v. 13, part 13, 47-2, p. 1-541, (2143).

Describes the mining and the economic geology of the gold and silver deposits of the U. S. A part of the Reports of the 10th Census.

King, T. Butler, 1850, Report on California, HED 59, v. 8, 31-1, p. 2-32, (577).

A report by a special government agent that describes the gold region and the mining.

Kocher, A. E., and **Hurst, Lewis A.,** 1907a, Soil survey of the Carrington area, North Dakota, HD 925, v. 110, 59-1, p. 927-948, 2 maps (in v. 111), (5050-5051).

Kocher, A. E., and **Hurst, Lewis A.,** 1907b, Soil survey of Sarpy County, Nebraska, HD 925, v. 110, 59-1, p. 893-909, 1 map (in v. 111), (5050-5051).

Kummel, Henry B., 1905, Geographical work of the New Jersey Geological Survey, HD 460, v. 111, 58-3, p. 752-756, (4890).

In the Report of the 8th International Geographic Congress.

Kunz, George, F., 1889, Gem collection of the U. S. National Museum, SMD 170, part 2, v. 12, 49-2, p. 267-275, (2499).

In the Annual Report of the U. S. National Museum.

Lacey, John F., 1904, Petrified Forest National Park of Arizona, HR 296, v. 1, 58-2, p. 1-8, (4577).

Describes the topography and scenic features of the area.

Lacey, John F., 1906, Petrified Forest National Park of Arizona, HR 4638, v. 3, 59-1, p. 1-8, (4908).

Describes the landforms and the geological features of the area and recommends the establishment of a National Park.

Lacroix, A., 1907, The eruption of Vesuvius in April, 1906, HD 575, v. 97, 59-2, p. 223-248, 13 plates, 1 map, (5200).

In the Annual Report of the Smithsonian Institution.

Laidley, T. T. S., 1856, Timber and mineral resources of the Deep River country, North Carolina, HED 109, v. 12, 34-1, p. 1-5, (859).

Describes the iron and coal deposits of the area.

Lalleman, G., 1905, Relations de la figure du globe avec la distribution des volcans et tremblements de terre, HD 460, v. 111, 58-3, p. 455-464, (4890).

Discusses the distribution of volcanoes and earthquakes on the earth. In the Report of the 8th International Geographic Congress.

Lander, Frederick W., 1855, Synopsis of a report of the reconnaissance of a railroad route from Puget Sound via south pass to the Mississippi, SED 78, v. 13, part 2, 33-2, p. 1-45, (759).

Also in HED 91, v. 11, part 2, 33-2, p. 1-45, (792). Published with the Reports of the Pacific Railroad Surveys, but not a result of those surveys.

Langfitt, W. C., 1903, Improvement of Willamette River, and of Columbia River below the mouth of the Willamette, and their tributaries, Oregon and Washington, HD 2, v. 12, 58-2, p. 2257-2321, 50 profiles and maps, (4638).

Describes the fluvial and shore features of the area.

Langford, N. P., 1873a, On the resources of Snake River Valley, HMD 112, v. 3, 42-3, p. 86-91, (1573).

Describes the mineral resources of parts of Idaho. In Hayden, F. V., 1873.

Langford, N. P., 1873b, Report of the superintendent of the Yellowstone National Park for the year 1872, SED 35, v. 1, 42-3, p. 1-9, (1545).

Includes a description of the topography and of the geysers and thermal springs.

Langley, S. P., 1902, The greatest flying creature, HD 707, part 1, v. 126, 57-1, p. 649-654, 3 plates, (4393).

Describes the Pterodactylia. In the Annual Report of the Smithsonian Institution.

Lapham, J. E., 1903, Soil survey of the Stuttgart area, Arkansas, HD 473, v. 106, 57-2, p. 611-622, 1 map, (4545).

Lapham, J. E., 1904, Soil survey of the Norfolk area, Virginia, HD 746, v. 111, 58-2, p. 233-252, 1 map (in v. 112), (4737-4738).

Lapham, J. E., and **Bennett, Hugh H.,** 1905, Soil survey of the Auburn area, New York, HD 458, v. 108, 58-3, p. 95-118, 1 map (in v. 109), (4887-4888).

Lapham, J. E., and **Lyman, W. S.,** 1907, Soil survey of Perquimans and Pasquotank Counties, North Carolina, HD 925, v. 110, 59-1, p. 271-288, 1 map (in v. 111), (5050-5051).

Lapham, J. E., Lyman, W. S., and **Ely, Charles W.,** 1907, Soil survey of Spalding County, Georgia, HD 925, v. 110, 59-1, p. 351-361, 1 map (in v. 111), (5050-5051).

Lapham, J. E., and **Meeker, F. N.,** 1904, Soil survey of the Asheville area, North Carolina, HD 746, v. 111, 58-2, p. 279-297, 1 map (in v. 112), (4737-4738).

Lapham, J. E., and **Miller, M. F.,** 1902, Soil survey of Montgomery County, Tennessee, HD 655, v. 117, 57-1, p. 341-357, 5 plates, 1 map, (4384).

Lapham, J. E., and **Mooney, Charles N.,** 1907a, Soil survey of the Cleveland area, Ohio, HD 925, v. 110, 59-1, p. 695-714, 1 map (in v. 111), (5050-5051).

Lapham, J. E., and **Mooney, Charles N.,** 1907b, Soil survey of the Westerville area, Ohio, HD 925, v. 110, 59-1, p. 715-729, 1 map (in v. 111), (5050-5051).

Lapham, J. E., and **Olshausen, B. A.,** 1903, Soil survey of the Wichita area, Kansas, HD 473, v. 106, 57-2, p. 623-642, 1 map, (4545).

Lapham, J. E., et al, 1903, Soil survey of the Vernon area, Texas, HD 473, v. 106, 57-2, p. 365-381, 1 map, (4545).

Lapham, J. E., et al, 1904, Soil survey of the Woodville area, Texas, HD 746, v. 111, 58-2, p. 511-520, 1 map (in v. 112), (4737-4738).

Lapham, Macy H., 1904, Soil survey of the San Jose area, California, HD 746, v. 111, 58-2, p. 1183-1217, 1 map (in v. 112), (4737-4738).

Lapham, Macy H., and **Ely, Charles W.,** 1907, Soil survey of the Gallatin Valley, Montana, HD 925, v. 110, 59-1, p. 975-996, 1 map (in v. 111), (5050-5051).

Lapham, Macy H., and **Heileman, W. H.,** 1902a, Soil survey of the Hanford area, California, HD 655, v. 177, 57-1, p. 447-480, 2 plates, 4 maps, (4384).

Lapham, Macy H., and **Heileman, W. H.,** 1902b, Soil survey of the Lower Salinas Valley, California, HD 655, v. 117, 57-1, p. 481-519, 2 plates, 3 maps, (4384).

Lapham, Macy H., and **Jensen, Charles A.,** 1905, Soil survey of the Bakersfield area, California, HD 458, v. 108, 58-3, p. 1089-1114, 3 maps (in v. 109), (4887-4888).

Lapham, Macy, H., and **Mackie, W. W.,** 1905, Soil survey of the Sacramento area, California, HD 458, v. 108, 58-3, p. 1049-1087, 1 map (in v. 109), (4887-4888).

Lapham, Macy H., and **Mackie, W. W.,** 1907, Soil survey of the Stockton area, California, HD 925, v. 110, 59-1, p. 997-1031, 2 maps (in v. 111), (5050-5051).

Lapham, Macy H., and **Neill, N. P.,** 1904, Soil survey of the Sollomonsville area, Arizona, HD 746, v. 111, 58-2, p. 1045-1070, 2 maps (in v. 112), (4737-4738).

Lapham, Macy H., et al, 1903, Soil survey of the Lower Arkansas Valley, Colorado, HD 473, v. 106, 57-2, p. 729-776, 3 plates, 2 maps, (4545).

Lapworth, Charles, 1904, The relations of geology, HD 748, v. 114, 58-2, p. 363-390, (4740).

Describes the relationship of geology to other sciences. In the Annual Report of the Smithsonian Institution.

LeConte, Joseph, 1885, Lectures on coal, SMD 272, v. 4, 35-1, p. 119-168, (937).

Also in HMD 135, v. 3, 35-1, p. 119-168, (963). Describes the origin and occurrence of coal, the climate of the "coal forming periods", and the structure and affinities of the coal forming plants. In the Annual Report of the Smithsonian Institution.

LeConte, Joseph, 1898, Earth-crust movements and their causes, HD 352, part 1, v. 72, 54-2, p. 233-244, (3548).

In the Annual Report of the Smithsonian Institution.

LeConte, Joseph, 1901, A century of geology, HD 537, part 1, v. 114, 56-2, p. 265-287, (4188).

Describes the advances in the science since 1800: In the Annual Report of the Smithsonian Institution.

Leidy, Joseph, 1852, Report upon some fossil mammalia and chelonia, from Nebraska, SMD 108, v. 1, 32-1, p. 63-65, (629).

Briefly describes fossils found by Thaddeus Culbertson and Stewart Van Vliet. In the Annual Report of the Smithsonian Institution.

Leidy, Joseph, 1872a, On the fossil vertebrates of the early Tertiary formations of Wyoming, HED 326, v. 15, 42-2, p. 353-372, (1520).

Part of Hayden, F. V., 1872a.

Leidy, Joseph, 1872b, Report on the vertebrate fossils of the Tertiary formations of the West, HED 325, v. 15, 42-2, p. 340-370, (1520).

A part of Hayden, F. V., 1872b.

Leith, Charles K., 1907, Iron ore reserves, HD 575, v. 97, 59-2, p. 207-214, (5200).

Describes the extent of the known reserves. In the Annual Report of the Smithsonian Institution.

Lesquereux, L., 1872a, Fossil flora: enumeration and description of the fossil plants . . . obtained (by) Dr. Hayden, 1870 and 1871: Remarks on the Cretaceous species: Tertiary flora of North America, HED 326, v. 15, 42-2, p. 283-318, (1520).

Part of Hayden, F. V., 1872a.

Lesquereux, L., 1872b, On the fossil plants of the Cretaceous and Tertiary formations of Kansas and Nebraska, HED 325, v. 15, 42-2, p. 370-385, (1520).

A part of Hayden, F. V., 1872b.

Lesquereux, L., 1873, Lignitic formation and fossil flora (of the Rocky Mountains), HMD 112, v. 3, 42-3, p. 315-427, (1573).

In Hayden, F. V., 1873.

Lieber, Oscar M., 1861, Notes on the geology of the coast of Labrador, HED 14, v. 7, 36-2, p. 402-408, 1 plate, (1098).

Describes the coastal geology and landforms of parts of Labrador and Newfoundland. In the Annual Report of the Coast Survey.

Lindenkohl, A., 1885, Geology of the sea bottom in the approaches to New York Bay, HED 43, v. 22, 48-2, p. 435-438, 1 map, (2297).

Describes the bottom topography and sediments of the area. In the Annual Report of the Coast and Geodetic Survey.

Linn, L. F., and **Sevier, A. H.,** 1836, Letters . . . relative to obstructions . . . of the White, Big Black, and St. Francis Rivers (Missouri and Arkansas), SED 113, v. 2, 24-1, p. 1-7, (280).

Describes natural log dams on the rivers, economic deposits of the area, and the topographic effects of the 1811 earthquake.

Linnard, T. B., 1844, Report on the improvement of the navigation of Red River, Louisiana, SED 1, v. 1, 28-2, p. 283-293, (449).

Describes the natural log dams on the river.

Little, George, 1882, Report on the blue clay of the Mississippi River, SED 12, v. 2, 46-3, p. 145-171, 1 map, (1942).

Describes the loess and fluvial deposits along the river. In the Annual Report of the Coast and Geodetic Survey.

Liveing, G. D., 1893, Crystallization, HMD 114, part 1, v. 22, 52-2, p. 269-280, (3131).

Describes the internal structure and the growth of crystals. In the Annual Report of the Smithsonian Institution.

Locke, John, 1844, (Geological report of . . .), SED 407, v. 7, 28-1, p. 147-189, 7 plates and maps, (437).

Also in HED 239, v. 6, 26-1, p. 116-159, (368), and also in HED 168, v. 4, part 2, 28-2, p. 116-159, 8 plates and maps, (467). Describes the geology of parts of Iowa, Wisconsin, and Illinois.

Locke, John, 1848, (Synopsis of a report on a survey of upper Michigan), SED 2, v. 2, 30-1, p. 183-199, (504).

A progress report describing the geology of the area covered during the last field season. In Jackson, C. T., 1848a.

Locke, John, 1849, Catalogue of specimens forewarded to Dr. Jackson . . ., SED 1, v. 3, 31-1, p. 563-572, (551).

A part of Jackson's final report. In Jackson, Charles T., 1849.

Lockwood, Daniel W., 1872, Report of (exploration through Nevada and Arizona), SED 65, v. 2, 42-2, p. 73-89, (1479).

Part of Wheeler, George M., 1872a.

Loew, Oscar, 1875, Geological and mineralogical report on portions of Colorado and New Mexico, HED 1, part 2, v. 5, 44-1, p. 1017-1036, (1676).

In Wheeler, George M., 1875.

Loew, Oscar, 1876a, Report on the geological and mineralogical character of southeastern California . . ., HED 1, part 2, v. 5, 44-2, p. 408-419, (1745).

In Wheeler, George M., 1876.

Loew, Oscar, 1876b, Report on the physical and agricultural features of Southern California, and especially the Mohave Desert, HED 1, part 2, v. 5, 44-2, p. 434-442, (1745).

In Wheeler, George M., 1876.

Long, S. H., 1827, Surveys of proposed routes of a national road from the city of Washington to Buffalo in the state of New York, HED 105, v. 5, 19-1, p. 7-39, 3 tables, (152).

Describes the topography and general lithology of the area surveyed.

Long, S. H., 1841, Report on the improvement of Red River, SED 64, v. 1, 27-1, p. 1-22, 1 map, (390).

Describes a natural log dam and associated fluvial features.

Long, S. H., Graham, R., and **Phillips, Joseph,** 1819, Topographical reports, made with a view to ascertain the practicability of uniting the waters of Illinois River, with those of Lake Michigan, HED 17, v. 2, 16-1, p. 5-10, (32).

Describes the topographic and fluvial features of the area.

Long, S. H., and **Humphreys, A. A.,** 1851, Report of the Board of Topographical Engineers on the inundations, etc., of the lower Mississippi, SED 13, v. 3, 31-2, p. 2-14, (589).

Describes the topographic and fluvial features of the area.

Loughridge, R. H., 1884a, Physical and agricultural description of the cotton region of Missouri, HMD 42, part 5, v. 13, part 5, 47-2, p. 5-15, (2133).

In the reports of the 10th Census. Describes the landforms and the soils of the state.

Loughridge, R. H., 1884b, Physio-geographical and agricultural features of the state of Arkansas, HMD 42, part 5, v. 13, part 5, 47-2, p. 9-39, 3 maps, (2133).

In the reports of the 10th Census. Describes the landforms and the soils of the state.

Loughridge, R. H., 1884c, Physio-geographical and agricultural features of the state of Texas, HMD 42, part 5, v. 13, part 5, 47-2, p. 11-57, 2 maps, (2133).

In the reports of the 10th Census. Describes the landforms and the soils of the state.

Loughridge, R. H., 1884d, Physio-geographical and agricultural features of the Indian Territory, HMD 42, part 5, v. 13, part 5, 47-2, p. 5-19, (2133).

In the reports of the 10th Census. Describes the landforms and the soils of parts of Oklahoma.

Lucas, Frederick A., 1901a, The restoration of extinct animals, HD 537, part 1, v. 114, 56-2, p. 479-492, 8 plates, (4188).

Describes the restoration of vertebrate fossils. In the Annual Report of the Smithsonian Institution.

Lucas, Frederick A., 1901b, The truth about the Mammoth, HD 737, part 1, v. 119, 56-1, p. 353-359, 3 plates, (4016).

Describes the morphology, distribution, and habits, of Mammoths. In the Annual Report of the Smithsonian Institution.

Lucas, Frederick A., 1902, The dinosaurs or terrible lizards, HD 707, part 1, v. 126, 57-1, p. 641-647, 4 plates, (4393).

Describes the main types of dinosaurs and the general features of the group. In the Annual Report of the Smithsonian Institution.

Ludlow, William, 1875, Report of a reconnaissance of the Black Hills of Dakota made in the summer of 1874, HED 1, part 2, v. 5, 44-1, p. 1113-1230, 1 plate, 3 maps, (1676).

Describes the exploration and general geology of the area. See also Winchell, N. H., 1875; and Grinell, George B., 1875.

Ludlow, William, 1876, Reconnaissance from Carroll, Montana, to the Yellowstone National Park, in the summer of 1875, HED 1, part 2, v. 5, 44-2, p. 570-699, 2 plates, 3 maps, (1745).

Describes the exploration and general geology of parts of Montana and Wyoming. See also Dana, Edward S., and Grinell, George B., 1876; and Whitfield, R. P., 1876.

Lull, E. P., 1879, Report of explorations and surveys for the location of a ship canal between the Atlantic and Pacific Ocean through the Isthmus of Panama, SED 75, v. 2, 45-3, p. 1-52, 4 maps and profiles, (1829).

Describes the topographic features of part of Panama. Bound in the same volume, but not part of, SED 13, the Annual Report of the Coast and Geodetic Survey.

Lydeckker, R., 1901, Mammoth ivory, HD 737, part 1, v. 119, 56-1, p. 361-366, (4016).

Describes the history of fossil ivory and the current trade in fossil ivory. In the Annual Report of the Smithsonian Institution.

Lyle, D. A., Report of (exploration through Nevada, Arizona, and parts of California), SED 65, v. 2, 42-2, p. 73-89, (1479).

In Wheeler, George M., 1872a.

Lyman, Benjamin S., 1869, The Freiberg School of Mines, HED 54, v. 9, 40-3, p. 230-238, (1374).

In Raymond, Rossiter W., 1869.

Lyman, W. S., Bennett, Frank, and **McLendon, W. E.,** 1908, Soil survey of Madison County, Tennessee, HD 352, v. 75, 59-2, p. 687-700, 1 map (in v. 76), (5178-5179).

Lynch, W. F., 1848, Report on an examination of the Dead Sea, SED 34, v. 4, 30-2, p. 1-88, 1 plate, (532).

Describes the topographic features of the area.

Maack, G. A., 1874, Report upon the geology and natural history of the Isthmus of Choco, of Darien, and of Panama, HMD 113, v. 5, 42-3, p. 155-175, 5 plates, (1575).

Describes the general geology of parts of Panama and Colombia. In Selfridge, Thomas O., 1874.

MacGonigle, John N., 1905, The Everglades of Florida, HD 460, v. 111, 58-3, p. 767-771, (4890).

Describes the paludal and lacustrine features. In the Report of the 8th International Geographic Congress.

Macomb, Alexander, Knight, Jonathan, and **Wevers, C. W.,** 1826, Report of the chief engineer in relation to the survey . . . of the road from the . . . Ohio . . . to the state of Missouri, HED 51, v. 4, 19-1, p. 1-32, (134).

Describes the topography, general lithology, and engineering geology of the routes surveyed.

Macomb, M. M., 1882a, Report . . . , for the fiscal year ending June 30, 1881, HED 1, part 2, v. 5, 47-1, p. 2809, 1 map, (2013).

A report, by an assistant, of the progress of the work under Wheeler's direction.

Macomb, M. M., 1882b, Report . . . for the fiscal year ending June 30, 1882, HED 1, part 2, v. 5, 47-2, p. 2821-2824, 1 map, (2094).

A report, by an assistant, of the progress of the work under Wheeler's direction.

Macomb, M. M., 1883, Report . . . for the fiscal year ending June 30, 1883, HED 1, part 2, v. 5, 48-1, p. 2379-2381, (2185).

A report, by an assistant, of the progress of the work under Wheeler's direction.

Macrae, Archibald, 1855, Report of a journey across the Andes and Pampas of the Argentine provinces, HED 121, v. 15, part 2, 33-1, p. 1-82, 2 maps, (729).

Describes the topography and general lithology of parts of Chile and Argentina. In v. 2 of the U. S. Astronomical Expedition to the Southern Hemisphere. See also Gilliss, J. M., 1855.

Mallet, R., 1860, On observation of earthquake phenomena, HMD 90, v. 7, 36-1, p. 408-433, (1066).

Describes earthquake waves and the primary and secondary effects of these waves and describes instruments for recording earthquakes. In the Annual Report of the Smithsonian Institution.

Mangum, A. W., and **Belden, H. L.,** 1905, Soil survey of the Austin area, Texas, HD 458, v. 108, 58-3, p. 421-446, 1 map (in v. 109), (4887-4888).

Mangum, A. W., and **Carr, M. Earl,** 1907, Soil survey of the Waco area, Texas, HD 925, v. 110, 59-1, p. 567-599, 1 map (in v. 111), (5050-5051).

Mangum, A. W., and **Drake, J. A.,** 1904, Soil survey of the Russell area, Kansas, HD 746, v. 111, 58-2, p. 911-926, 1 map (in v. 112), (4737-4738).

Mangum, A. W., and **Lee, Ora, Jr.,** 1908, Soil survey of the Laredo area, Texas, HD 352, v. 75, 59-2, p. 481-504, 1 map (in v. 76), (5178-5179).

Mangum, A. W., and **Lyman, W. S.,** 1908, Soil survey of the San Marcos area, Texas, HD 352, v. 75, 59-2, p. 505-537, 1 map (in v. 76), (5178-5179).

Mangum, A. W., and **Mann, Charles J.,** 1905, Soil survey of the Owosso area, Michigan, HD 458, v. 108, 58-3, p. 665-687, 1 map (in v. 109), (4887-4888).

Mangum, A. W., and **Neill, N. P.,** 1905a, Soil survey of the Boonville area, Indiana, HD 458, v. 108, 58-3, p. 727-749, 1 map (in v. 109), (4887-4888).

Mangum, A. W., and **Root, Aldert S.,** 1904, Soil survey of the Campobello Indiana, HD 458, v. 108, 58-3, p. 707-726, 1 map (in v. 109), (4887-4888).

Mangum, A. W., and **Root, Aldert S.,** 1904, Soil survey of the Campobello area, South Carolina, HD 746, v. 111, 58-2, p. 299-315, 1 map (in v. 112), (4737-4738).

Mangum, A. W., and **Schroeder, F. C.,** 1908, Soil survey of the Crookston area, Minnesota, HD 352, v. 75, 59-2, p. 865-891, 1 map (in v. 76), (5178-5179).

Mann, Charles J., and **Tharp, W. E.,** 1908, Soil survey of Putnam County, Missouri, HD 352, v. 75, 59-2, p. 893-910, 1 map (in v. 76), (5178-5179).

Manson, Marsden, 1902, Features and water rights of Yuba River, California, SD 356, v. 27, 57-1, p. 115-130, 1 plate, 1 map, (4246).

Describes the landforms and fluvial features of the area. See also Nutter, Edward H., 1902; and Soulé, Frank, 1902.

Marbut, C. F., 1905, Physiography in the university, HD 460, v. 111, 58-3, p. 997-1004, (4890).

Discusses the place of geomorphology in the college curriculum. In the Report of the 8th International Geographic Congress.

Marcou, John B., 1885, A review of the progress of North American invertebrate paleontology for 1884, SMD 33, part 1, v. 2, 48-2, p. 563-582, (2266).

Includes a bibliography for the year 1884. In the Annual Report of the Smithsonian Institution.

Marcou, John B., 1886, Record of North American invertebrate paleontology (for 1885), HMD 15, v. 25, 49-1, p. 713-759, (2431).

Includes a bibliography for 1885. In the Annual Report of the Smithsonian Institution.

Marcou, John B., 1889, North American invertebrate paleontology for 1886, HMD 600, part 1, v. 17, 50-1, p. 231-229, (2581).

Includes a bibliography for 1886. In the Annual Report of the Smithsonian Institution.

Marcou, Jules, 1855a, Field notes (of geology of the route near the 35th parallel), SED 78, v. 13, part 3, 33-2, p. 121-164, (760).

Also in HED 91, v. 11, part 3, 33-2, p. 121-164, (793). A bilingual, French and English, publication of field notes. In Whipple, A. W., 1855, a part of the Pacific Railroad Survey Reports.

Marcou, Jules, 1855b, Geological notes of a survey of the country comprised between Preston, Red River, and El Paso, Rio Grande del Norte, HED 129, v. 18, part 2, 33-1, p. 125-128, (737).

A report based on field notes sent to Marcou. In Pope, John, 1855, but only in the version published during the 1st session of the 33rd Congress.

Marcou, Jules, 1855c, Resume of a geological reconnaissance (of a route near the 35th parallel), SED 78, v. 13, part 3, 33-2, p. 165-171, (760).

Also in HED 129, v. 18, part 2, 33-1, p. 40-48, (737) and in HED 91, v. 11, part 3, 33-2, p. 165-171, (793). In Whipple, A. W., 1855, forms part of the Pacific Railroad Survey Reports.

Marcou, Jules, 1876, Report on the geology of a portion of Southern California, HED 1, part 2, v. 5, 44-2, p. 378-392, (1745).

In Wheeler, George M., 1876.

Marcy, Randolph B., 1850, Report (on a route from Fort Smith to Santa Fe), SED 64, v. 14, 31-1, p. 169-233, (562).

Describes the landforms and rocks of parts of Arkansas, Texas, and New Mexico.

Marcy, Randolph B., 1853, Exploration of the Red River of Louisiana in the year 1852, SED 54, v. 8, 32-2, p. 1-319, 61 plates, 2 maps, (666).

A detailed journal of exploration that contains descriptions of topography and lithology. See also Shepard, Charles U., 1853; and Shumard, George 1853a, 1853b; and Hitchock, Edward, 1853.

Marcy, Randolph B., 1856, Report of an expedition to the sources of the Brazos and Big Witchita Rivers, during the summer of 1854, SED 60, v. 12, 34-1, p. 2-47, 1 map. (821).

Describes the exploration, lithology, and topography of parts of Texas. See also Blake, W. P., 1856a.

Marsh, O. C., 1878, Letter transmitting . . . the report on the scientific surveys of the territories . . ., SMD 9, v. 1, 45-3, p. 1-5, (1833).

A report by the National Academy of Sciences urging the establishment of a United States Geological Survey. See also Marsh, O. C., 1879a, and 1879b.

Marsh, O. C., 1879a, Letter . . . transmitting a report on the surveys of the territories, HMD 5, v.1, 45-3, p. 1-27, (1861).

The same as Marsh, O. C., 1878, but containing additional letters. See also Marsh, O. C., 1879b.

Marsh, O. C., 1879b, Report on surveys of the territories, HMD 7, v. 1, 46-1, p. 19-22, (1876).

See also Marsh, O. C., 1878 and 1879a.

Marean, Herbert W., 1902, Soil survey of the Covington area, Georgia, HD 655, v. 117, 57-1, p. 329-340, 3 plates, 1 map, (4384).

Marean, Herbert W., 1903a, Soil survey of Posey County, Indiana, HD 473, v. 106, 57-2, p. 441-463, 2 plates, 1 map, (4545).

Marean, Herbert W., 1903b, Soil survey of Union County, Kentucky, HD 473, v. 106, 57-2, p. 425-440, 1 map, (4545).

Marean, Herbert W., and **Jones, Grove B.**, 1904a, Soil survey of Cerro Gordo County, Iowa, HD 746, v. 111, 58-2, p. 853-873, 1 map (in v. 112), (4737-4738).

Marean, Herbert W., and **Jones, Grove B.**, 1904b, Soil survey of Story County, Iowa, HD 746, v. 111, 58-2, p. 833-851, 1 map (in v. 112), (4737-4738).

Marinden, Henry L., 1882, Comparison of the surveys of Delaware River in front of Philadelphia, 1843 and 1878, SED 12, v. 2, 46-3, p. 110-125, 4 maps, (1942).

Describes the channel geometry and the changes in fluvial features of this part of the river. In the Annual Report of the Coast and Geodetic Survey.

Marinden, Henry L., 1883, Comparison of the survey of Delaware River of 1819, between Pettys and Tinicum Islands, with more recent surveys, SED 77, v. 4, 47-2, p. 427-432, 3 plates, (2077).

Describes the changes in fluvial features in parts of Delaware, New Jersey, and Pennsylvania. In the Annual Report of the Coast and Geodetic Survey.

Marinden, Henry L., 1885, Physical hydrography of Delaware River and Bay. Comparison of recent with former surveys, HED 43, v. 22, 48-2, p. 431-434, 2 maps, (2297).

Describes the changes in fluvial features in parts of Delaware, New Jersey, and Pennsylvania. In the Annual Report of the Coast and Geodetic Survey.

Marinden, Henry L., 1890a, Cross-sections of the shore of Cape Cod between Chatham and the Highland light house, HED 55, v. 27, 51-1, p. 409-457, (2742).

Describes the shore features of this part of Massachusetts. In the Annual Report of the Coast and Geodetic Survey.

Marinden, Henry L., 1890b, Encroachment of the sea upon the coast of Cape Cod, Massachusetts, as shown by comparitive surveys, HED 55, v. 27, 51-1, p. 403-407, (2742).

Describes the shoreline changes in the area. In the Annual Report of the Coast and Geodetic Survey.

Marinden, Henry L., 1892a, Cross-sections of the shore of Cape Cod, Massachusetts, between the Cape Cod and Long Point light-houses, HED 43, part 2, v. 32, 52-1, p. 289-341, 1 map, (2952).

Describes the shore features of the area. In the Annual Report of the Coast and Geodetic Survey.

Marinden, Henry L., 1892b, On the changes in the shore lines and anchorage areas of Cape Cod (or Provincetown) Harbor as shown by a comparison of surveys made between 1835, 1867, and 1890, HED 43, part 2, v. 32, 52-1, p. 283-288, 1 map, (2952).

Describes the shore features and their changes. In the Annual Report of the Coast and Geodetic Survey.

Marinden, Henry L., 1894, On the changes in the ocean shore lines of Nantucket Island, Massachusetts, from a comparison of surveys made in the years 1846 to 1887 and in 1891, SED 37, part 2, v. 4, 52-2, p. 243-252, 4 maps, (3058).

Describes the shore features and their changes. In the Annual Report of the Coast and Geodetic Survey.

Marinden, Henry L., 1896, Report on the changes in the depths on the bar at the entrance to Nantucket inner harbor, Massachusetts, between the years 1888 and 1893, SD 25, v. 2, 54-1, p. 347-354, 4 plates and maps, (3348).

Describes the changes in shore and shallow water features. In the Annual Report of the Coast and Geodetic Survey.

Marinden, Henry L., 1897, Tables of cross-sections on the north shores of Nantucket and Marthas Vineyard, Massachusetts, SD 35, v. 2, 54-2, p. 305-346, (3468).

Describes the shore features and their changes. In the Annual Report of the Coast and Geodetic Survey.

Martel, E. A., 1905, Scientific exploration of caves, HD 460, v. 111, 58-3, p. 165-172, 4 plates, (4890).

Describes the features of caves found in the United States. In the Report of the 8th International Geographic Congress.

Martin, J. O., 1902, Soil survey of the Willis area, Montgomery Co., Texas, HD 655, v. 117, 57-1, p. 607-619, 1 map, (4384).

Martin, J. O., 1904, Soil survey of the Lockhaven area, Pennsylvania, HD 746, v. 111, 58-2, p. 129-142, 1 map (in v. 112), (4737-4738).

Martin, J. O., and **Ayrs, O. L.**, 1905, Soil survey of the Jackson area, Mississippi, HD 458, v. 108, 58-3, p. 343-352, 1 map (in v. 109), (4887-4888).

Martin, J. O., and **Carr, E. P.**, 1904a, Soil survey of the Ashtabula area, Ohio, HD 746, v. 111, 58-2, p. 647-658, 1 map (in v. 112), (4737-4738).

Martin, J. O., and **Carr, E. P.**, 1904b, Soil survey of Miller County, Arkansas, HD 746, v. 111, 58-2, p. 563-576, 1 map (in v. 112), (4737-4738).

Martin, J. O., and **Sweet, A. T.**, 1905, Soil survey of the Kearney area, Nebraska, HD 458, v. 108, 58-3, p. 859-874, 3 maps (2 in v. 109), (4887-4888).

Marvine, A. R., 1871, Notes on reported mineral deposits lying between Santo Domingo City and Azua, etc., SED 9, v. 1, p. 105-111, (1466).

Includes a brief description of the areal geology of this part of the Dominican Republic.

Matteucci, Carlo, 1871, On the electrical currents of the earth, HED 153, v. 12, 41-3, p. 208-255, (1460).

Discusses the electrical resistivity of rocks and structures of the crust. In the Annual Report of the Smithsonian Institution.

Matthes, Francois E., 1905, The Lewis Range of northern Montana and its glaciers, HD 460, v. 111, 58-3, p. 478-479, (4890).

A brief abstract in the Report of the 8th International Geographic Congress.

McCaskey, H. D., 1905a, Report of the Chief of the Mining Bureau, HD 2, v. 12, 59-1, p. 291-328, (4952).

Describes the known mineral resources of the Philippines.

McCaskey, H. D., 1905b, Report of the Mining Bureau, HD 2, v. 13, 58-3, p. 379-410, 9 plates, (4792).

Describes the mineral resources of the Philippines and the organization of the Philippine Mining Bureau.

McCaskey, H. D., 1907, Mining Bulletin No. 3: Report on a geological reconnaissance of the iron region of Angat, Bulacan, HD 2, v. 10, 59-2, p. 715-749, (5113).

Describes the landforms, springs, and mineral deposits of this part of the Philippines.

McCauley, C. A. H., 1878, Report on the San Juan reconnaissance of 1877 . . ., HED 1, part 2, v. 5, 45-3, p. 1750-1871, 7 plates, 2 maps, (1846).

Describes the landforms and mineral resources of parts of Colorado and New Mexico.

McCauley, C. A. H., 1879, Notes on Pagosa Springs, Colorado, SED 65. v. 4, 45-3, p. 3-27, (1831).

Describes the landforms and mineral springs of the area.

McClellan, George B., 1871, Report on the Dominican Republic in the year 1854, HED 43, v. 7, 41-3, p. 1-8, (1453).

Includes a brief description of the topography and the coal mines.

McCrary, George W., 1879, Information relative to the surveys in the territories west of the Mississippi River . . ., SED 21, part 2, v. 1, 45-3, p. 1-17 and p. 1-6, (1828).

Letters relating to the establishment of a national geological Survey.

McGee, W. J., 1890, Geology for 1887 and 1888, HMD 142, part 1, v. 14, 50-2, p. 217-260, (2668).

In the Annual Report of the Smithsonian Institution.

McGee, W. J., et al, 1893, Geology of Washington and vicinity, HMD 107, v. 13, 53-2, p. 219-252, (3241).

A field guide to the geology of parts of Virginia and Maryland. In Congrès Geologique International, 1893.

McIntyre, H. H., 1870, (Report on Alaska), SED 32, v. 1, 41-2, p. 25-42, (1405).

Includes a brief description of the topography and mineral resources of the coastal regions.

McIntyre, James, 1849, Report of . . ., SED 1, v. 3, 31-1, p. 506-509, (551).

Also in HED 5, v. 3, part 3, 31-1, p. 506-509, (571). Describes the geology of part of the Lake Superior region.

McLendon, W. E., 1904, Soil survey of the Blackfoot area, Idaho, HD 746, v. 111, 58-2, p. 1027-1044, 2 maps (in v. 112), (4737-4738).

McLendon, W. E., and Carr, M. Earl, 1905, Soil survey of the Saginaw area, Michigan, HD 458, v. 108, 58-3, p. 603-638, 1 map (in v. 109), (4887-4888).

McLendon, W. E., and Jones, Grove B., 1908, Soil survey of Oklahoma County, Oklahoma, HD 352, v. 75, 59-2, p. 563-585, 1 map (in v. 76), (5178-5179).

McLendon, W. E., and **Lyman, W. S.,** 1908, Soil survey of Grainger County, Tennessee, HD 352, v. 75, 59-2, p. 661-686, 1 map (in v. 76), (5178-5179).

McLendon, W. E., and **Mann, Charles J.,** 1907, Soil survey of Montgomery County, Alabama, HD 925, v. 110, 59-1, p. 425-452, 1 map (in v. 111), (5050-5051).

Means, Thomas H., 1900a, A reconnoissance in the Cache a la poudre Valley, Colorado, HD 399, v. 88, 56-1, p. 121-124, (3985).

A reconnaissance study of the soils of the area by the Soil Survey.

Means, Thomas H., 1900b, A reconnoissance in Sanpete, Cache, and Utah Counties, Utah, HD 399, v. 88, 56-1, p. 115-120, (3985).

A reconnaissance study of the soils of the area by the Soil Survey.

Means, Thomas H., 1901, Soil survey in Salt River Valley, Arizona, HD 526, v. 107, 56-2, p. 287-332, 4 plates, 3 maps, (4181).

Means, Thomas H., and **Gardner, Frank D.,** 1900, A soil survey in the Pecos Valley, New Mexico, HD 311, v. 88, 56-1, p. 36-76, 9 plates, 6 maps, (3985).

Means, Thomas H., and **Holmes, J. Garnett,** 1901, Soil survey around Fresno, Cal., HD 526, v. 107, 56-2, p. 333-384, 16 plates, 2 maps, (4181).

Means, Thomas H., and **Holmes, J. Garnett,** 1902, Soil survey around Imperial, California, HD 655, v. 117, 57-1, p. 587-606, 9 plates, 2 maps, (4384).

Meek, F. B., 1871, Report on the paleontology of eastern Nebraska, HED 19, v. 2, 42-1, p. 83-239, 11 plates, (1471).

Describes the invertebrate fossils. In Hayden, F. V., 1871.

Meek, F. B., 1872a, (Instructions for the study of geology on the expedition to the North Pole), SMD 149, v. 2, 42-2, p. 381-384, (1482).

Also in HED 1, part 3, v. 4, 42-2, p. 255-258, (1507). In the Annual Report of the Smithsonian Institution.

Meek, F. B., 1872b, Preliminary list of the fossils collected by Dr. Hayden's exploring expedition of 1871, in Utah, and Wyoming Territories, with descriptions of a few new species, HED 326, v. 15, 42-2, p. 373-377, (1520).

Part of Hayden, F. V., 1872a.

Meek, F. B., 1872c, Preliminary paleontological report (on the Hayden survey of Wyoming etc.), HED 325, v. 15, 42-2, p. 287-318, (1520).

In Hayden, F. V., 1872b.

Meek, F. B., 1873, Paleontological report (in connection with the Hayden Survey), HMD 112, v. 3, 42-3, p. 429-518, (1573).

Describes invertebrates collected in Idaho, Montana, Utah, and Wyoming. In Hayden, F. V., 1873.

Meeker, F. N., and **Avon-Burke, R. T.,** 1907, Soil survey of Portage County, Wisconsin, HD 925, v. 110, 59-1, p. 837-864, 1 map (in v. 111), (5050-5051).

Meeker, F. N., and **Tailby, G. W., Jr.,** 1908, Soil survey of Meigs County, Ohio, HD 352, v. 75, 59-2, p. 701-728, 1 map (in v. 76), (5178-5179).

Mendell, G. H., 1882, Report upon a project to protect the navigable waters of California from the effects of hydraulic mining, HED 98, v. 20, 47-1, p. 3-110, 3 maps, (2028).

Describes the landforms and fluvial features of the placer regions and the environmental changes caused by hydraulic mining.

Mendenhall, Walter C., 1898a, The Alaska peninsula and the Aleutian Islands, HD 172, v. 13, 55-3, p. 115-117, (3737).

Describes the general geology and the mineral deposits.

Mendenhall, Walter C., 1898b, The Kadiak Islands, HD 172, v. 13, 55-3, p. 113-114, (3737).

Describes the general geology and the mineral deposits of this part of Alaska.

Mendenhall, Walter C., 1898c, The Kenai peninsula, HD 172, v. 13, 55-3, p. 109-110, (3737).

Describes the general geology and the mineral deposits of this part of Alaska.

Mendenhall, Walter C., 1899, Report on the region between Resurrection Bay and the Tanana River, SD 172, v. 13, 55-3, p. 40-50, (3737).

Describes the general geology and the mineral deposits of this part of Alaska.

Mendenhall, Walter C., 1901, A reconnaissance in the Norton Bay region, Alaska, in 1900, HD 547, v. 124, 56-2, p. 181-222, 3 plates, 3 maps, (4198).

Describes the general geology and the gold deposits.

Merino, Miguel, 1864, Figure of the earth, HMD 83, v. 4, 38-1, p. 306-330, (1201).

Describes the various methods used in determining the shape of the earth. In the Annual Report of the Smithsonian Institution.

Merriam, C. Hart, 1902, Bogoslof volcanoes, HD 707, part 1, v. 126, 57-1, p. 367-375, 2 plates, (4393).

Describes the volcanic activity of the area. In the Annual Report of the Smithsonian Institution.

Merrill, George P., 1889, The collection of building and ornamental stones in the U. S. National Museum: handbook and catalogue, SMD 170, part 2, v. 12, 49-2, p. 277-648, 9 plates, (2499).

In the Annual Report of the U. S. National Museum.

Merrill, George P., 1890, Petrography for 1887 and 1888, HMD 142, part 1, v. 14, 50-2, p. 327-354, (2668).

Includes a bibliography for 1887 and 1888. In the Annual Report of the Smithsonian Institution.

Merrill, George P., 1891, Handbook for the department of geology in the U. S. National Museum. Part 1—Geognosy—The materials of the earth's crust, HMD 129, part 2, v. 11, 51-2, p. 503-591, 12 plates, (2879).

Describes the classification and physical properties of the rocks in the collections. In the Annual Report of the U. S. National Museum.

Merrill, George P., 1897, Notes on the geology and natural history of the peninsula of Lower California, HD 25, part 2, v. 82, 54-1, p. 969-994, 9 plates, 1 map, (3449).

Describes the landforms and general geology of Baja California and Baja California Sur. In the Annual Report of the U. S. National Museum.

Merrill, George P., 1901, Guide to the study of the collections in the section of applied geology: the nonmetallic minerals. HD 737, part 2, v. 120, 56-1, p. 155-485, 30 plates, (4017).

In the Annual Report of the U. S. National Museum.

Merrill, George P., 1906, Contributions to the history of American geology, HD 430, v. 107, 58-3, p. 189-733, 37 plates, 1 map, (4886).

In the Annual Report of the U. S. National Museum.

Mesmer, Louis, 1903, Soil survey of the Lewiston area, Idaho, HD 473, v. 106, 57-2, p. 689-709, 1 map, (4545).

Mesmer, Louis, 1904, Soil survey of the Los Angeles area, California, HD 746, v. 111, 58-2, p. 1263-1306, 2 maps (in v. 112), (4737-4738).

Mesmer, Louis, and **Hearn, W. E.,** 1903, Soil survey of the Bigflats area, New York, HD 473, v. 106, 57-2, p. 125-142, 2 plates, 1 map, (4545).

Michler, N., 1857a, (Report on the area) from the mouth of Devils River to El Paso del Norte, SED 108, v. 20, part 1, 34-1, p. 74-80, (832).

Also in HED 135, v. 14, part 1, 34-1, p. 74-80, (861). Describes the topography and lithology of part of the area of the U. S. Mexican Boundary Survey. In Emory, William H., 1857.

Michler, N., 1857b, (Report on the area) from the 111th meridian of longitude to the Pacific Ocean, SED 108, v. 20, part 1, 34-1, p. 101-136, (832).

Also in HED 135, v. 14, part 1, 34-1, p. 101-136, (861). Describes the topography and lithology of part of the area of the U. S. Mexican Boundary Survey. In Emory, William H., 1857.

Michler, N., 1861, (Report of survey for an interoceanic canal near the Isthmus of Darien), SED 9, v. 7, part 1 and 2, 36-2, p. 3-457, 17 maps and profiles (in serial number 1086), (1085-1086).

Describes the landforms and general geology of parts of Colombia. See also Schott, Arthur, 1861a and 1861b.

Mitchell, Henry, 1861, Description of implements devised . . . for collecting specimens of bottom in alluvial harbors, HED 14, v. 7, 36-2, p. 398, 1 plate, (1098).

Describes the construction of a bottom sediment sampling device. In the Annual Report of the Coast Survey.

Mitchell, Henry, 1874, Report . . . concerning Nausett Beach and the peninsula of Monomy, HED 121, v. 11, 42-2, p. 134-143, (1514).

Describes the shore features and the beach changes of this part of Massachusetts. In the Annual Report of the Coast Survey.

Mitchell, Henry, 1875, Additional report concerning the changes in the neighboorhood of Chatham and Monomy, HED 133, v. 11, 43-1, p. 103-107, 1 plate, (1609).

Describes the changes in certain Massachusetts beaches. In the Annual Report of the Coast Survey.

Mitchell, Henry, 1877, Report on an inspection of the terminal parts of the proposed canals through Nicaragua and the Isthmus of Darien, HED 100, v. 14, 43-2, p. 135-147, (1647).

Describes the coastal features of parts of Panama and Nicaragua.

Mitchell, Henry, 1880, Notes concerning alleged changes in the relative elevations of land and sea, SED 12, v. 4, 45-2, p. 98-103, (1783).

A discussion opposed to the alleged emergence of the New England Coast. In the Annual Report of the Coast Survey.

Mitchell, Henry, 1881, On a physical survey of the Delaware River in front of Philadelphia, SED 13, v. 2, 45-3, p. 121-173, (1829).

Describes the channel geometry and other fluvial features. In the Annual Report of the Coast and Geodetic Survey.

Mitchell, Henry, 1883, Study of the effect of River bends in the lower Mississippi, SED 77, v. 4, 47-2, p. 433-436, 1 plate, (2077).

Describes the channel geometry and other fluvial features of this part of the river. In the Annual Report of the Coast and Geodetic Survey.

Mitchell, Henry, 1884, The estuary of the Delaware, SED 29, v. 3, 48-1, p. 239-245, 1 map, (2164).

Describes the fluvial and coastal features of parts of Delaware, New Jersey, and Pennsylvania. In the Annual Report of the Coast and Geodetic Survey.

Mitchell, Henry, 1887a, A report on Monomy and its shoals, HED 40, v. 22, 49-2, p. 255-260, 1 map, (2481).

Describes the coastal features of parts of Massachusetts. In the Annual Report of the Coast and Geodetic Survey.

Mitchell, Henry, 1887b, A report on the delta of the Delaware, HED 40, v. 22, 49-2, p. 267-279, (2481).

Describes the fluvial and coastal features of parts of Delaware and New Jersey. In the Annual Report of the Coast and Geodetic Survey.

Mitchell, Henry, 1889a, Addendum to a report on the estuary of the Delaware, HED 17, v. 24, 50-1, p. 269-273, (2556).

Describes the fluvial and coastal features of the area. In the Annual Report of the Coast and Geodetic Survey.

Mitchell, Henry, 1889b, On the movements of the sand at the eastern entrance to Vineyard Sound, HED 17, v. 24, 50-1, p. 159-163, (2556).

Describes the changes in beach features in parts of Massachusetts. In the Annual Report of the Coast and Geodetic Survey.

Moe, Alfred K., 1904, Honduras: geographical sketch, natural resources . . ., HD 145, part 4, v. 66, 58-3, p. 1-252, 15 plates, 2 maps, (4845).

Describes the landforms and mineral resources of the country.

Mooney, Charles N., and **Ayrs, O. L.,** 1905a, Soil survey of the Greeneville area, Tennessee-North Carolina, HD 458, v. 108, 58-3, p. 493-525, 1 map (in v. 109), (4887-4888).

Mooney, Charles N., and **Ayrs, O. L.,** 1905b, Soil survey of Lawrence County, Tennessee, HD 458, v. 108, 58-3, p. 475-492, 1 map (in v. 109), (4887-4888).

Mooney, Charles N., and **Bonsteel, F. E.,** 1903, Soil survey of the Albermarle area, Virginia, HD 473, v. 106, 57-2, p. 187-238, 3 plates, 3 maps, (4545).

Mooney, Charles N., and **Caine, Thomas A.,** 1902, Soil survey of the Prince Edward area, Virginia, HD 655, v. 117, 57-1, p. 259-271, 1 map, (4384).

Mooney, Charles N., Martin, F. O., and **Caine, Thomas A.,** 1902, Soil survey of the Bedford area, Virginia, HD 655, v. 117, 57-1, p. 239-257, 6 plates, 1 map, (4384).

Mooney, Charles N., Westover, H. L., and **Bennett, Frank,** 1908, Soil survey of Merrimack County, New Hampshire, HD 352, v. 75, 59-2, 1 map (in v. 76), (5178-5179).

Mooney, Charles N., et al, 1907, Soil survey of Lavaca County, Texas, HD 925, v. 110, 59-1, p. 623-642, 1 map (in v. 111), (5050-5051).

Morris, William G., 1879, Report upon the customs district, public services, and resources of Alaska Territory, SED 59, v. 4, 45-3, p. 1-163, 1 map, (1831).

Includes descriptions of the known coal deposits and other economic deposits.

Mullan, John, 1863, Construction of a military road from Fort Walla Walla to Fort Benton, SED 43, v. 1, 37-3, p. 2-363, 8 plates, 4 maps, (1149).

Describes the landforms and topography from the Missouri River to Walla Walla, Washington. The fossils collected were later described by Meek & Hayden the Proc. Philadelphia Acad. Nat. Sci.

Murray, John, 1901, Present condition of the floor of the ocean; evolution of the continental and oceanic areas, HD 737, part 1, v. 119, 56-1, p. 309-328, (4016).

Describes abyssal sediments and features and discusses the origin of ocean basins. In the Annual Report of the Smithsonian Institution.

Murray, John, 1905, Deep sea deposits, HD 460, v. 111, 58-3, p. 407, (4890).

An abstract in the Report of the 8th International Geographic Congress.

Neill, N. P., and **Tharp, W. E.,** 1907a, Soil survey of Newton County, Indiana, HD 925, v. 110, 59-1, p. 747-770, 1 map (in v. 111), (5050-5051).

Neill, N. P., and **Tharp, W. E.,** 1907b, Soil survey of Tippecanoe County, Indiana, HD 925, v. 110, 59-1, p. 781-813, 1 map (in v. 111), (5050-5051).

Neill, N. P., et al, 1904, Soil survey of the Laramie area, Wyoming, HD 746, v. 111, 58-2, p. 1071-1097, 3 maps (in v. 112), (4737-4738).

Nettleton, C. E., 1892, Final report on artesian and underflow investigation, SED 41, part 2, v. 4, 52-1, p. 1-116, 27 plates and maps, (2899).

Describes the ground water conditions and resources in parts of the Great Plains and Rocky Mountains.

Newberry, J. S., 1855, Report upon the geology of the route (from Sacramento Valley to the Columbia River), SED 78, v. 13, part 6, 33-2, p. 1-85, (763).

Also in HED 91, v. 11, part 6, 33-2, p. 1-85, (796). Describes the general geology of parts of California and Oregon. Part of the Pacific Railroad Survey Reports.

Newberry, J. S., 1861, Geological report (for the Colorado River exploring expedition under Lt. Ives), HED 90, v. 14, 36-1, p. 1-154, 6 plates, (1058).

Describes the geology of the lower Colorado River area in parts of California, Arizona, and Nevada. In Ives, Joseph C., 1861.

Newberry, J. S., 1872, The ancient lakes of western America; their deposits and drainage, HED 325, v. 15, 42-2, p. 329-339, (1520).

Describes the glacial and paleolacustrine features of parts of the Rocky Mountains, Great Plains, and Great Basin areas. In Hayden, F. V., 1872b.

Nicaraguan Canal Commission, 1906, Documents relating to the interoceanic canal and . . . the earthquake at Panama in 1882, SD 264, v. 6, 59-1, p. 1-73, (4914).

Describes and discusses the earthquake in relation to the construction of the canal.

Nicollet, I. N., 1843, Report intended to illustrate a map of the hydrographical basin of the Upper Mississippi River, SED 237, v. 5, part 2, 26-2, p. 1-170, 1 map, (380).

Also in HED 52, v. 2, 28-1, p. 1-170, 1 map, (different from the above) (464). One of the earliest detailed descriptions of the geology of this area. J. C. Fremont assisted Nicollet. A very important work.

Nordenskiold, Otto, et al, 1904, The Swedish Antarctic expedition, HD 748, v. 114, 58-2, p. 467-479, 1 map, (4740).

Describes the general geology and the landforms of parts of Antarctica. In the Annual Report of the Smithsonian Institution.

Norris P. W., 1877, Report on the Yellowstone National Park, HED 1, part 5, v. 8, 45-2, p. 837-845, (1800).

Describes the topography, thermal springs, and geysers of the area.

Norwood, J. G., 1848, (Report on the geology of parts of Wisconsin and the Lake Superior region), SED 57, v. 7, 30-1, p. 73-129, (509).

A geological report by one of Owen's assistants. In Owen, David D., 1848b.

Nutter, Edward H., 1902, Water bearing gravels and formations tributary to the underground water supply of the Salinas Valley in Monterey County, SD 356, v. 27, 57-1, p. 208-213, (4246).

Describes the hydrogeology of the area. See also Manson, Marsden, 1902; and Soulé, Frank, 1902.

Obermaier, Hugues, 1906, Quaternary human remains in Central Europe, HD 575, v. 97, 59-2, p. 373-397, (5200).

In the Annual Report of the Smithsonian Institution.

Ortmann, A. E., 1905, Origin of the deep-sea fauna, HD 460, v. 111, 58-3, p. 618-620, (4890).

In the Report of the 8th International Geographic Congress.

Orton, Edward, 1893, Origin of the rock pressure of natural gas in the Trenton limestone of Ohio and Indiana, HMD 334, part 1, v. 43, 52-1, p. 127-153, (3001).

Describes the geology and the natural gas resources of parts of Indiana and Ohio. In the Annual Report of the Smithsonian Institution.

Owen, David D., 1840, Report of a geological exploration of part of Iowa, Wisconsin, and Illinois, HED 239, v. 6, 26-1, p. 10-161, 25 plates, (368).

Also in HED 168, v. 4, part 2, 28-2, p. 1-161, 27 plates and maps, (467); and also in SED 407, v. 7, 28-1, p. 1-191, 25 plates and maps, (437). The first detailed geologic description of the area. Plates and maps are found in (437) and (467). For (368) there were two issues. In the first, spelled Wiskonsin, there were no plates, in the second, some volumes have plates and some lack them.

Owen, David D., 1844, (Letter on the coal formations of Indiana and adjoining states), SED 78, v. 2, 28-1, p. 8-9, (432).

A letter written to support a petition for the completion of the Wabash and Erie Canal.

Owen, David D., 1848a, Preliminary report containing outlines of the progress of the geological survey of Wisconsin and Iowa, SED 2, v. 2, 30-1, p. 160-173, (504).

A progress report covering areas not examined in his earlier surveys.

Owen, David D., 1848b, Report of a geological reconnoissance of the Chippewa land district of Wisconsin and the northern part of Iowa, SED 57, v. 7, 30-1, p. 5-134, 37 plates, 1 map, (509).

Covers the same general areas as Owen, David D., 1848a, but includes more information based on additional field work.

Owen, David D., 1852, (Letter concerning the exploration of the "bad lands"), SED 1, v. 1, 32-2, p. 81-84, (658).

A letter recommending an extension of geological surveys into Nebraska and South Dakota.

Owen, David D., 1868, Artesian wells (an extract from a geological reconnoissance of Arkansas in 1859-60), HED 273, v. 17, 40-2, p. 32-34, (1343).

A brief description of the known wells. In Taylor, James W., 1868.

Packard, A. S., 1886a, On the Anthracaridae, family of Carboniferous macrurous Crustacea, SMD 154, v. 7, 49-1, p. 135-139, 2 plates, (2348).

In the Memoirs of the National Academy of Sciences—v. 3, part 2, mem. 15, part 3.

Packard, A. S., 1886b, On the Carboniferous Xiphosurous fauna of North America, SMD 154, v. 7, 49-1, p. 143-157. 4 plates, (2348).

In the Memoirs of the National Academy of Sciences—v. 3, part 2, mem. 16.

Packard, A. S., 1886c, On the Gampsyonychidae, an undescribed family of Schizopod Crustacea, SMD 154, v. 7, 49-1, p. 129-133, 2 plates, (2348).

In the Memoirs of the National Academy of Sciences—v. 3, part 2, mem. 15, part 2.

Packard, A. S., 1886d, On the Syncarida, a hitherto undescribed synthetic group of extinct Malacostracous Crustacea, SMD 154, v. 7, 49-1, p. 121-128, 2 plates, (2348).

In the Memoirs of the National Academy of Sciences—v. 3, part 2, mem. 15, part 1.

Packard, A. S., 1905, Evidence in favor of the former connection of Brazil and Africa, and of an originally antarcticogaeic land mass, HD 460, v. 111, 58-3, p. 638-640, (4890).

In the Report of the 8th International Geographic Congress.

Palmer, Aaron H., 1848, Memoir, geographical, political, and commercial, on . . . Siberia, Manchuria, and the Asiatic Islands of the Northern Pacific Ocean, SMD 80, v. 1, 30-1, p. 1-77, 1 map, (511).

> Includes landform description and a description of the known Siberian mineral deposits.

Palmieri, L., 1871, The electro-magnetic seismograph, HED 20, v. 2, 42-1, p. 425-428, (1471).

> Describes a new type of seismograph. In the Annual Report of the Smithsonian Institution.

Parke, John G., 1855a, Report of exploration for railroad routes from San Francisco Bay to Los Angeles, California . . ., and from the Pimas villages on the Gila to the Rio Grande . . ., SED 78, v. 13, part 7, 33-2, p. 1-42, 8 plates, 2 maps, (764).

> Also in HED 91, v. 11, part 7, 33-2, p. 1-42, 8 plates, 2 maps, (797). Part of the Pacific Railroad Survey Reports, see also Antisell, Thomas, 1855.

Parke, John G., 1855b, Report of explorations for that portion of a railroad route near the 32nd parallel of north latitude, lying between Dona Ana, on the Rio Grande, and the Pimas villages, on the Gila, SED 78, v.13, part 2, 33-2, p. 1-28, (759).

> Also in HED 91, v. 11, part 2, 33-2, p. 1-28, (792). Part of the Pacific Railroad Survey Reports.

Parry, C. C., 1869, (Report on the Colorado River), HED 1, part 2, v. 5, 40-3, p. 1191-1195, (1368).

> Describes the supposed journey of James White through the canyon and describes the geology as indicated by White's account.

Parry, C. C., and Schott, Arthur, 1857, Geological reports (to accompany the United States and Mexican Boundary Survey), additional notes by W. H. Emory, SED 108, v. 20, part 1, 34-1, p. 1-174, 21 plates, (832).

> Also in HED 135, v. 14, part 1, 34-1, p. 1-174, 21 plates, (861). Describes in detail the geology of the boundary area. See also Emory, William H., 1857.

Paxton, Joseph, 1829, Letter (concerning the) Raft of Red River, SED 78, v. 1, 20-2, p. 1-18, (181).

> Describes the natural log dam and the associated fluvial features, and includes a discussion on the origin and growth of the dam.

Peale, A. C., 1872, Report . . . on minerals, rocks, thermal springs, etc., HED 326, v. 15, 42-2, p. 165-204, (1520).

> Describes the geology of parts of Montana, Wyoming, Utah, and Idaho. In Hayden, F. V., 1872a.

Peale, A. C., 1873, (Geological report on Colorado, Utah, Montana, and Wyoming), HMD 112, v. 3, 42-3, p. 99-189, (1573).

In Hayden, F. V., 1873.

Peale, A. C., 1883, The thermal springs of Yellowstone National Park, HMD 62, part 2, v. 22, 47-1, p. 63-454, (2057).

Describes and compares the region with other geyser regions of the world. In Hayden, F. V., 1883b.

Peckham, S. F., 1884, Production, technology, and uses of petroleum and its products, HMD 42, part 10, v. 13, part 10, 47-2, p. 1-301, 30 plates, 11 maps and charts, (2139).

A complete description that includes history, origin, geographic distribution and geologic occurrence of petroleum. A part of the reports of the 10th census.

Peirce, Benjamin, 1881, On the internal constitution of the earth, SED 17, v. 2, 46-2, p. 201, (1883).

A brief mathematical description of the increase in density with depth. In the Annual Report of the Coast and Geodetic Survey.

Penck, Albrecht, 1905, The valleys and lakes of the Alps, HD 460, v. 111, 58-3, p. 173-184, (4890).

Describes the fluvial, lacustrine, and glacial features of the region. In the Report of the 8th International Geographic Congress.

Perry, M. C., 1856, Narrative of the expedition of an American squadron to the China Seas and Japan performed in the years 1852, 1853, and 1854, HED 97, v. 12, part 1, 33-2, p. 1-537, 89 plates, 1 map, (802).

Also in SED 79, v. 14, part 2, 33-2, p. 1-537, 89 plates, 1 map, (769). A narrative compiled by F. L. Hanks from Perry's notes. Includes brief hydrographic, topographic, and lithologic descriptions.

Peters, W. J., and **Brooks, Alfred H.,** 1899, Report on the White River-Tanana expedition, SD 172, v. 13, 55-3 p. 64-75, (3737).

Describes the geology and the economic deposits of this part of Alaska.

Pfizenmayer, E., 1907, A contribution to the morphology of the mammoth *Elephas primigenius* Blumenbach; with an explanation of my attempt at a restoration, HD 575, v. 97, 59-2, p. 321-333, 1 plate, (5200).

In the Annual Report of the Smithsonian Institution.

Pfordte, Otto F., 1905, The glaciers of Poto, Peru, HD 460, v. 111, 58-3, p. 497-500, (4890).

In the Report of the 8th International Geographic Congress.

Philippi, R. A., 1855, Meteoric iron of Atacama, HED 121, v. 15, part 2, 33-1, p. 287-289, (729).

A brief description of meteorites found in Chile. In v. 2 of the U. S. Astronomical Expedition to the Southern Hemisphere. See also Gilliss, J. M., 1855.

Piersin, William M., 1874, Correspondence relative to the discovery of a large meteorite in Mexico, SMD 130, v. 2, 43-1, p. 419-422, (1585).

Describes a meteorite found in Chihuahua. In the Annual Report of the Smithsonian Institution.

Pilar, George, 1877, The revolutions of the crust of the earth, SMD 46, v. 3, 44-2, p. 283-357, (1724).

Discusses the origin of the earth, the central heat of the earth, sediments and their origin, and glaciation. In the Annual Report of the Smithsonian Institution.

Pope, John, 1850, Report of an exploration of the territory of Minnesota, SED 42, v. 10, 31-1, p. 2-56, 1 map, (558).

Describes the landforms and some of the lithology of the area.

Pope, John, 1855, Report of exploration of a route for the Pacific railroad near the 32nd parallel of latitude from the Red River to the Rio Grande, SED 78, v. 13, part 2, 33-2, p. 1-186, 10 plates, (759).

Also in HED 91, v. 11, part 2, 33-2, p. 1-186, 10 plates, (792); also in HED 129, v. 18, part 2, 33-1, p. 1-324, (737). Part of the Pacific Railroad Survey Reports, this describes the southern route. Blake, William P., 1855, was supposed to form part of this report, but was instead published separately. Marcou, Jules, 1855b, which is a preliminary geological report on this route, is omitted here.

Pope, John, 1858, (Letters relating to the) Artesian well experiment, SED 1, v. 2, 35-2, p. 590-608, (975).

Describes efforts to drill wells near the Pecos River in Texas. Includes brief descriptions of the rocks encountered in drilling.

Pope, John, 1860, Report on an artesian experiment and on the geology of the region near Galisteo, New Mexico, SED 2, v. 2, 36-1, p. 544-549, (1024).

Contains a brief description of the general geology and a detailed description of the drilling and its results.

Pourtales, L. F., 1854, Extracts of letters on the examination of specimens of bottom obtained in the exploration of the Gulf Stream, SED 14, v. 13, 33-1, p. 82-83, (704).

Also in HED 12, v. 4, 33-1, p. 82-83, (716). A brief description of deposits whose turbidity current origin was not recognized at the time. In the Annual Report of the Coast Survey.

Pourtales, L. F., 1859, Microscopical examination of specimens of bottom from deep sea soundings, SED 14, v. 16, 35-2, p. 248-250, (991).

> Also in HED 33, v. 6, 35-2, p. 248-250, (1005). A review of studies on samples collected in the Atlantic and Caribbean portions of the Gulf Stream. It contains a description of sequential changes in foraminifera tests during the formation of what seems to be glauconite. In the Annual Report of the Coast Survey.

Pourtales, L. F., 1871, Report on dredgings made in the sea near the Florida reefs, HED 71, v. 11, 40-3, p. 168-170, (1379).

> Briefly describes the sediments and the organisms found. In the Annual Report of the Coast Survey.

Pourtales, L. F., 1872, The Gulf Stream—characteristics of the Atlantic sea bottom off the coast of the United States, HED 206, v. 8, 41-2, p. 220-225, (1419).

> Describes the bottom sediments. The map referred to in the text is not included. In the Annual Report of the Coast Survey.

Powell, Charles F., 1905, Improvement of rivers and harbors in Connecticut, and of Pawcatuck River, Rhode Island and Connecticut, HD 2, v. 6, 59-1, p. 871-952, (4946).

> Describes the fluvial features and the sediments of the Connecticut River.

Powell, John W., 1872a, Exploration of the Colorado River and its tributaries explored in 1869, 1870, 1871, and 1872, HMD 300, v. 6, 43-1, 81 plates, 1 map, (1622).

> Powell's classic work and the final report of the explorations.

Powell, John W., 1872b, Report preliminary for continuing the survey of the Colorado of the west and its tributaries, HMD 173, v. 3, 42-2, p. 2-12, (1562).

> A progress report on the exploration of the river.

Powell, John W., 1873, (Second preliminary) report of the survey of the Colorado of the west, SMD 76, v. 2, 42-3, p. 2-16, (1572).

> A progress report on geological and ethnological work under Powell's direction.

Powell, John W., 1877, Report on the geographical and geological survey of the Rocky Mountain Region, HED 1, part 5, v. 8, 45-2, p. 789-805, (1800).

> A progress report describing the geology of parts of Arizona, Nevada, South Dakota, and Utah.

Powell, John W., 1878a, (Letter regarding geographical and geological surveys), HED 80, v. 17, 45-2, p. 1-19, 1 map, (1809).

A letter describing Powell's work of the preceeding 10 years. See also Humphreys, A. A., 1878b; and Hayden, F. V., 1878a.

Powell, John W., 1878b, Report on the lands of the arid regions of the United States, with a more detailed account of the lands of Utah, HED 73, v. 13, 45-2, p. 1-195, 3 maps.

Describes the landforms of the arid regions in Arizona, California, Colorado, Nevada, New Mexico, Texas, and Utah.

Powell, John W., 1879, Letter . . . relating to the cost of geographical surveys, HED 72, v. 16, 45-3, p. 1-6, (1858).

Describes the cost of work under Powell's direction and urges the elimination of duplicate surveys under various government departments.

Powell, John W., 1886a, (Report on geological specimens collected by the Nicaragua surveying expedition, in 1885), SED 99, v. 5, 49-1, p. 49-51, (2337).

A letter describing various igneous and metamorphic rocks collected by the expedition.

Powell, John W., 1886b, (Report on the proposed construction of a railroad to Alaska), SMD 22, v. 1, 49-2, p. 2-10, 2 maps, (2450).

Describes what is known about the geology of the possible routes.

Preston, E. B., 1905, North Fork mining district of Fresno County, SD 34, v. 2, 58-3, p. 49-50, (4764).

Describes the geology and mineral resources of this part of California. See also Brown, Edwin C., 1905.

Prestwich, Joseph, 1876, The past and future of geology, SMD 115, v. 1, 44-1, p. 175-195, (1665).

Discusses invertebrate paleontology, the evolution of the earth, and the history of geology. In the Annual Report of the Smithsonian Institution.

Pumpelly, Raphael, 1886, Report on the mining industries of the United States, (exclusive of precious metals), with special invesigations into the iron resources of the Republic and into the Cretaceous coals of the northwest, HMD 42, part 15, v. 13, part 15, 47-2, p. 1-1025, 102 plates, (2145).

A part of the reports of the 10th Census.

Quale, Paul, 1867, An account of the Cryolite of Greenland, HMD 83, v. 1, 39-2, p. 398-401, (1302).

Describes the Fluorspar deposits and mines in southern Greenland. In the Annual Report of the Smithsonian Institution.

Raymond, Charles W., 1871, Report of a Reconnaissance of the Yukon River, Alaska, SED 12, v. 1, 42-1, p. 1-113, (1466).

Includes topographic description and brief descriptions of the mineral resources.

Raymond, Rossiter W., 1869, Mineral resources of the United States and territories west of the Rocky Mountains, HED 54, v. 9, 40-3, p. 3-256, (1374).

A continuation of the series started by Browne, J. R., 1867.

Raymond, Rossiter W., 1870, Statistics of mines and mining in the states and territories west of the Rocky Mountains, HED 207, v. 10, 41-2, p. 1-805, (1424).

Describes the mines and the economic geology of the region.

Raymond, Rossiter W., 1871, Mining statistics west of the Rocky Mountains, HED 10, v. 1, 41-3, p. 1-566, (1470).

Describes the mines and the economic geology of the region.

Raymond, Rossiter W., 1872, Statistics of mines and mining in the states and territories west of the Rocky Mountains, HED 211, v. 10, 42-2, p. 1-566, 6 plates, (1513).

Describes the mines and ecomomic geology of the region and contains sections on American schools of mining and on mining law.

Raymond, Rossiter W., 1873, Statistics of mines and mining in the states and territories west of the Rocky Mountains, HED 210, v. 9, 42-2, p. 1-550, 31 plates, 3 maps, (1567).

Describes the mines and the economic geology of the region. See also Bowman, Amos, 1873; and Waldeyer, Charles, 1873.

Raymond, Rossiter W., 1874, Statistics of mines and mining in the states and territories west of the Rocky Mountains, HED 141, v. 10, 43-1, p. 1-585, (1608).

Describes the mines and the economic geology of the region.

Raymond, Rossiter W., 1875, Statistics of mines and mining in the states and territories west of the Rocky Mountains, HED 177, v. 18, 43-2, p. 1-540, 4 plates, 3 maps, (1651).

Describes the mines and the economic geology of the region. See also Bowman, Amos, 1875.

Raymond, Rossiter W., 1877, Statistics of mines and mining in the states and territories west of the Rocky Mountains, HED 159, v. 14, 44-1, p. 1-519, 1 map, (1691).

Describes the mines and the economic geology of the region.

Reid, Harry F., 1892, Report of an expedition to Muir Glacier . . ., HED 43, part 2, v. 32, 52-1, p. 487-501, 1 map, (2952).

Describes the glacier and the adjacent landforms. In the Annual Report of the Coast and Geodetic Survey.

Reid, Harry F., 1905a, The glaciers of Mount Hood and Mount Adams, HD 460, v. 111, 58-3, p. 492, (4890).

An abstract in the Report of the 8th International Geographic Congress.

Reid, Harry F., 1905b, The reservoir lag in glacier variations, HD 460, v. 111, 58-3, p. 487-491, (4890).

Describes the movement of glacial ice. In the Report of the 8th International Geographic Congress.

Rhees, William J., 1901, The Smithsonian Institution: documents relative to its origin and history, HD 732, v. 113, parts 1 and 2, 56-1, p. 1-1983, 2 maps, (4010, v. 1 and 2).

Rice, Thomas D., 1904, Soil survey of Ouachita Parish, Louisiana, HD 746, v. 111, 58-2, p. 419-438, 1 map (in v. 112), (4737-4738).

Rice, Thomas D., 1907, Soil survey of McCracken County, Kentucky, HD 925, v. 110, 59-1, p. 679-694, 1 map (in v. 111), (5050-5051).

Rice, Thomas D., 1908, Soil survey of the Williston area, North Dakota, HD 352, v. 75, 59-2, p. 999-1022, 1 map (in v. 76), (5178-5179).

Rice, Thomas D., and **Ayrs, Orla L.,** 1908, Soil survey of the Tishomingo area, Indian Territory, HD 352, v. 75, 59-2, p. 539-562, 1 map (in v. 76), (5178-5179).

Describes the soils of part of Oklahoma.

Rice, Thomas D., and **Geib, W. J.,** 1905a, Soil survey of Coshocton County, Ohio, HD 458, v. 108, 58-3, p. 565-580, 1 map (in v. 109), (4887-4888).

Rice, Thomas D., and **Geib, W. J.,** 1905b, Soil survey of the Gainesville area, Florida, HD 458, v. 108, 58-3, p. 269-289, 1 map (in v. 109), (4887-4888).

Rice, Thomas D., and **Geib, W. J.,** 1905c, Soil survey of the Munising area, Michigan, HD 458, v. 108, 58-3, p. 581-601, 1 map (in v. 109), (4887-4888).

Rice, Thomas D., and **Geib, W. J.,** 1905d, Soil survey of Warren County, Kentucky, HD 458, v. 108, 58-3, p. 527-541, 1 map (in v. 109), (4887-4888).

Rice, Thomas D., and **Griswold, Lewis,** 1904a, Soil survey of Acadia Parish, Louisiana, HD 746, v. 111, 58-2, p. 461-485, 1 map (in v. 112), (4737-4738).

Rice, Thomas D., and **Griswold, Lewis,** 1904b, Soil survey of the New Orleans area, Louisiana, HD 746, v. 111, 58-2, p. 439-459, 1 map (in v. 112), (4737-4738).

Rice, Thomas D., and **Taylor, F. W.,** 1903, Soil survey of the Darlington area, South Carolina, HD 473, v. 106, 57-2, p. 291-307, 4 plates, 1 map, (4545).

Rice, William N., 1905, The classification of mountains, HD 460, v. 111, 58-3, p. 185-190, (4890).

In the Report of the 8th International Geographic Congress.

Rittenhouse, H. O., Knight, A. M., and **Huse, H. P.,** 1881, Report upon earthquake at Scio, April 3, 1881, HED 1, part 3, v. 8, 47-1, p. 746-752, 1 map, (2016).

Describes the effects of this earthquake in Turkey.

Rockwood, Charles G., Jr., 1885, Record of recent scientific progress, 1884 in Vulcanology and Seismology, SMD 33, part 1, v. 2, 48-2, p. 215-235, (2266).

Includes a bibliography for 1884. In the Annual Report of the Smithsonian Institution.

Rockwood, Charles G., Jr., 1886, Record of scientific progress, 1885 in Vulcanology and Seismology, HMD 15, v. 25, 49-1, p. 471-493, (2431).

Includes a bibliography for 1885. In the Annual Report of the Smithsonian Institution.

Rockwood, Charles G., Jr., 1889, Vulcanology and Seismology for 1886, HMD 600, part 1, v. 17, 50-1, p. 289-312, (2581).

Includes a bibliography for 1886. In the Annual Report of the Smithsonian Institution.

Roessler, A. R., Wilson, Joseph S., and **Goodwin, J. R.,** 1871, Report on the history and the present condition of the cabinet of minerals and natural history connected with the general land office, HED 69, v. 8, 41-3, p. 1-9, (1454).

A brief description of the minerals and some "new" mineral deposits.

Root, Aldert S., and **Hurst, Lewis A.,** 1905, Soil survey of Lancaster County, South Carolina, HD 458, v. 108, 58-3, p. 169-184, 1 map (in v. 109), (4887-4888).

Root, Aldert S., and **Hurst, Lewis A.,** 1907, Soil survey of Duplin County, North Carolina, HD 925, v. 110, 59-1, p. 289-307, 1 map (in v. 111), (5050-5051).

Ruffner, E. H., 1874, A report and map of a reconnaissance in the Ute country, made in 1873 . . . , HED 193, v. 12, 43-1, p. 1-101, (1610).

Describes the exploration and general geology of parts of Colorado and New Mexico. See also Campbell, Donald, 1874; Hawn, F., 1874a, 1874b, 1874c, 1974d; and Hawn, Laurens, 1874.

Ruffner, E. H., 1877a, Survey of the headwaters of Red River, Texas, HED 1, part 2, v. 4, 45-2, p. 1401-1438, (1796).

Describes the exploration and some of the geology of parts of Texas and Oklahoma.

Ruffner, E. H., 1877b, Geological notes (on explorations in the Department of Missouri), HED 1, part 2, v. 4, 45-2, p. 1431-1438, (1796).

Describes the strata and fossils of part of the Texas panhandle.

Russell, Israel C., 1903, Volcanic eruptions on Martinique and St. Vincent, HD 484, part 1, v. 109, 57-2, p. 331-349, 11 plates, (4548).

Describes the eruption of Mount Pelée that destroyed St. Pierre. In the Annual Report of the Smithsonian Institution. See also Anderson, Tempest and Flett, John S., 1903.

Rutimeyer, L., 1862, The fauna of middle Europe during the stone-age, SMD 77, v. 1, 37-2, p. 361-367, (1141).

Describes the Pleistocene vertebrate remains. In the Annual Report of the Smithsonian Institution.

Safford, James M., 1884, Physio-geographical and agricultural features of the state of Tennessee, HMD 42, part 5, v. 13, part 5, 47-2, p. 9-45, 2 maps, (2133).

In the reports of the 10th Census. Describes the landforms and the soils of the state.

St. John, Orestes H., 1871, Descriptions of fossil fishes from the Upper Coal-Measures of Nebraska, HED 19, v. 2, 42-1, p. 239-245, 3 plates, (1471).

In Hayden, F. V., 1871.

St. John, Orestes H., 1883, Geology of the Wind River District, HMD 62, part 1, v. 21, 47-1, p. 173-269, 49 plates, (2056).

In Hayden, F. V., 1883a.

Sanchez, Alfred M., 1904, Soil survey of the Provo area, Utah, HD 746, v. 111, 58-2, p. 1121-1150, 8 maps (6 in v. 112), (4737-4738).

Sanchez, Alfred M., 1905, (Report of the soil physicist of the Philippine Islands), HD 2, v. 13, 58-3, p. 521-531, (4792).

Describes the soils and their mechanical and chemical analysis.

Sands, B. F., 1856, . . . an instrument for procuring specimens of bottom in sounding, SED 22, v. 17, 34-1, p. 361, 1 plate, (826).

Also in HED 6, v. 3, 34-1, p. 361, 1 plate, (845). In the Annual Report of the Coast Survey.

Sapper, Karl, 1905, Grundzüge des Gebirgsbaus von Mittelamerika, HD 460, v. 111, 58-3, p. 231-238, 1 map, (4890).

Describes the volcanology and the structural geology of Central America. In the Report of the 8th International Geographic Congress.

Sartorius, Charles, 1867, The earthquake in eastern Mexico of the second of January, 1866, HMD 83, v. 1, 39-2, p. 432-434, (1302).

A brief eyewitness description of an earthquake in the province of Vera Cruz. In the Annual Report of the Smithsonian Institution.

Sartorius, Charles, 1871, Eruption of the volcano of Colima in June, 1869, HED 153, v. 12, 41-3, p. 422-423, (1460).

A brief description of the eruption of this Mexican volcano. In the Annual Report of the Smithsonian Institution.

Schaeffer, George C., 1855, Descriptions of the structure of the fossil wood from the Colorado Desert, SED 78, v. 13, part 5, 33-2, p. 338-339, 1 plate, (762).

Also in HED 91, v. 11, part 5, 33-2, p. 338-339, 1 plate, (795). Part of the Pacific Railroad Survey Reports. In Williamson, R. S., 1855.

Schiel, James, 1855a, Geological report of the country explored under the 38th and 41st parallels of north latitude, in 1853-'54, SED 78, v. 13, part 2, 33-2, p. 96-107.

Also in HED 91, v. 11, part 2, 33-2, p. 96-107, (792); and also in HED 129, v. 18, part 2, 33-1, p. 120-133, (737). Part of the Pacific Railroad Survey Reports. In Beckwith, E. G., 1855a.

Schiel, James, 1855b, List and description of the organic remains collected during the exploration of the central pacific railroad line . . . in 1853-'54, SED 78, v. 13, part 2, 33-2, p. 108-111, 4 plates, (759).

Also in HED 91, v. 11, part 2, 33-2, p. 108-111, 4 plates, (792), and also in HED 129, v. 18, part 2, 33-1, p. 134-135, (737). Part of the Pacific Railroad Survey Reports. In Beckwith, E. G., 1855a.

Schoolcraft, Henry R., 1822, Report on the number, value, and position of the copper mines on the southern shore of Lake Superior, SED 5, v. 1, 17-2, p. 7-28, (73).

Describes the strata and the location of collected copper samples. A stratigraphic section was included in the original report but is not included in this Senate document.

Schott, Arthur, 1861a, Remarks on the geognostic structure of the country, with accompanying descriptive table of geologic specimens, SED 9, v. 7, part 1, 36-2, p. 169-174, (1085).

Describes the structural geology of parts of Colombia. See also Schott, Arthur, 1861b; and Michler, N., 1861.

Schott, Arthur, 1861b, Report on the physiography of the Isthmus of Choco, SED 9, v. 7, part 1 and 2, p. 148-168, 1 profile, (1085-1086).

Describes the landforms of this part of Colombia. See also Schott, Arthur, 1861a; and Michler, N., 1861.

Schott, Charles C., 1875, On underground temperatures, HMD 56, v. 2, 43-2, p. 249-253, (1654).

Discusses crustal temperatures and their variation. In the Annual Report of the Smithsonian Institution.

Schrader, Frank C., 1898, The Prince William Sound and Copper River country, SD 172, v. 13, 55-3, p. 105-108, (3737).

Describes the geology and the economic deposits of this part of Alaska.

Schrader, Frank C., 1899, Report on Prince William Sound and the Copper River region, SD 172, v. 13, 55-3, p. 51-63, (3737).

Describes the geology and the economic deposits of this part of Alaska.

Schrader, Frank C., and **Brooks, Alfred H.,** 1900, Preliminary report on the Cape Nome Gold Region Alaska, SD 236, v. 25, 56-1, p. 1-56, 19 plates, 3 maps, (3867).

Describes the economic geology of the area.

Schrader, Frank C., and **Spencer, Arthur C.,** 1901, The geology and mineral resources of a portion of the Copper River district, Alaska, HD 546, v. 124, 56-2, p. 1-94, 9 plates, 5 maps, (4198).

Describes the geology and the copper and gold deposits of the area.

Schuchert, Charles, 1905, Karl Alfred von Zittel, HD 430, v. 106, 58-3, p. 779-786, 1 plate, (4885).

Describes the life and work of von Zittel. In the Annual Report of the Smithsonian Institution.

Schwatka, Frederick, 1884, Military reconnaissance in Alaska, SED 2, v. 1, 48-2, p. 3-121, 20 maps, (2261).

Describes the landforms from Chilkoot Inlet to Fort Selkirk on the Yukon River.

Scudder, Samuel H., 1885, Description of an Articulate of doubtful relationship from the Tertiary beds of Florissant, Colorado, SMD 69, v. 5, 48-2, p. 87-90, (2269).

Describes the remains of an insect. In the Memoirs of the National Academy of Sciences (v. 3, part 1, mem. 6).

Scudder, Samuel H., 1883, The Tertiary lake basin at Florissant, Colorado, HMD 62, part 1, v. 21, 47-1, p. 271-293, 1 map, (2056).

Describes the geology and fossils of the Florissant beds. In Hayden, F. V., 1883a.

Selfridge, Thomas O., 1871, Report on the survey of the Isthmus of Darien, 1871, HED 1, part 3, v. 4, 42-2, p. 178-203, 1 plate, 1 map, (1507).

Describes a reconnaissance in Panama and Colombia for a canal site. Includes a short section on geological features.

Selfridge, Thomas O., 1874, Report of explorations and surveys to ascertain the practicability of a ship canal . . . by way of the Isthmus of Darien, HMD 113, v. 5, 42-3, p. 1-268, 14 plates, 17 maps, (1574).

Describes the exploration and landforms of parts of Panama and Colombia. See also Bowditch, E. W., 1874; Carson, J. Petigru, 1874; and Maack, G. A., 1874.

Senate Committee on Military Affairs, 1900, Compilation of narratives of explorations in Alaska, SR 1023, v. 11, 56-1, p. 1-856, 25 plates, 25 maps, (3896).

Describes the areal geology and mineral resources of various parts of Alaska.

Senate Select Committee, 1838, (Report to accompany) a bill . . . to occupy the Oregon Territory, SED 470, v. 5, 25-2, p. 1-23, 2 maps, (318).

Includes an extract from the *Encyclopedia of Geography* that describes the topography and geology of the area.

Shaler, N. S., 1873, On the phosphate beds of South Carolina, HED 112, v. 11, 41-3, p. 182-189, (1459).

Includes a description of the geologic history of the state as well as a description of the fossiliferous phosphate beds. In the Annual Report of the Coast Survey.

Shaler, N. S., 1875, Geological resources of the Big Sandy Valley, HED 1, part 2, v. 4, 44-1, p. 763-765, (1675).

Describes the coal deposits of this part of Kentucky.

Shepard, Charles U., 1853, Report on the minerals collected (during an exploration of Red River), SED 54, v. 8, 32-2, p. 155-159, (666).

Describes the minerals and rocks collected along the river in Texas and Oklahoma.

Sheridan, Jo E., 1903, Annual report of the Mine Inspector for the Territory of New Mexico, HD 5, v. 21, 58-2, p. 743-818, 4 plates, 1 map, (4647).

Also in HD 5, v. 23, 58-2, p. 387-462, 1 plate, (4649). Describes the coal deposits of parts of the state.

Sheridan, Jo E., 1904, Report of the Mine Inspector for New Mexico, HD 5, v. 21, 58-3, p. 641-717, (4800).

Describes the coal resources and the mines of the area.

Shreve, Henry M., 1833, (Letter on removal of parts of the Red River Raft), SED 1, v. 1, 23-1, p. 126-127, (238).

Describes a natural log dam and associated fluvial features.

Shreve, Henry M., and **Gratiot, C.,** 1834, (Letters on clearing away the Raft of Red River), HED 98, v. 3, 23-1, p. 1-13, 1 map, (256).

Describes the log dam and the associated fluvial features. Letter 13 in this is the same as Shreve, Henry M., 1833.

Shufeldt, Robert W., 1872, Reports of explorations and surveys to ascertain the practicability of a ship canal . . . by way of the Isthmus of Tehuantepec, SED 6, v. 3, 42-2, p. 1-151, 11 plates, 20 charts and maps, (1480).

Describes the exploration and landforms of parts of Vera Cruz and Oaxaca. See also Fuertes, E. A., 1872; and Spear, John C., 1872.

Shumard, B. F., 1853a, Description of the species of Carboniferous and Cretaceous fossils collected (during an exploration of the Red River), SED 54, v. 8, 32-2, p. 199-211, 7 plates, (666).

Describes the invertebrates collected in parts of Texas and Oklahoma. See also Shumard, B. F., 1853b.

Shumard, B. F., 1853b, Remarks upon the general geology of the country traversed (during an exploration of the Red River), SED 54, v. 8, 32-2, p. 179-195, 9 plates, (666).

Describes the geology of parts of Texas and Oklahoma. See also Shumard, B. F., 1853a.

Shumard, B. F., 1858, Paleontology (of a geological exploration from Fort Leavenworth to Bryan's Pass), SED 11, v. 3, 35-1, p. 517-520, (920).

Also in HED 2, v. 2, part 2, 35-1, p. 517-520, (943). Briefly describes fossils collected on the Great Plains. See also Engelmann, Henry, 1858.

Siguel, F., 1871, Tour to the petroleum wells, to Azua Viejo, to the "Agua Hediondo" . . . , SED 9, v. 1, 42-1, p. 103-105, (1466).

Includes a brief description of the geology of this part of the Dominican Republic.

Simpson, J. H., 1850a, . . . report and map of the route from Fort Smith to Santa Fe, New Mexico, HED 12, v. 6, 31-1, p. 2-25, 1 map, (554).

Describes the topography and rocks of parts of Arkansas, Texas, and New Mexico.

Simpson, J. H., 1850b, Report on an expedition into the Navajo country in 1849, SED 64, v. 14, 34-1, p. 56-168, 75 plates, 2 maps, (562).

Includes descriptions of the landforms and the lithology of parts of New Mexico and Arizona.

Simpson, J. H., 1859, Report and map of wagon road routes in Utah Territory, SED 40, v. 10, 35-2, p. 2-84, 1 map, (984).

Includes descriptions of the landforms and lithology of parts of Utah and Wyoming. See also Engleman, Henry, 1859. In 1876 an expanded version was published by the Corps of Engineers, but not as a Congressional document.

Simpson, J. H., and **Marcy, R. B.,** 1850, Report on a route from Fort Smith to Santa Fe, HED 45, v. 8, 31-1, p. 2-89, 1 map, (577).

Describes the landforms and lithology of parts of Arkansas, Texas, and New Mexico.

Sitgreaves, L., 1853, Report of an expedition down the Zuni and Colorado Rivers, SED 59, v. 10, 32-2, p. 4-198, 78 plates, 1 map, (668).

Describes the landforms and lithology of parts of Arizona, California, and Nevada.

Smart, Charles, 1879, Chemical report on the Pagosa Springs of Colorado, HED 1, part 2, v. 5, 46-2, p. 2332-2344, (1906).

Describes the mineral springs and waters of the area.

Smith, Eugene A., 1884a, Physio-geographical and agricultural features of the state of Georgia, HMD 42, part 6, v. 13, part 6, 47-2, p. 9-67, 5 plates and maps, (2134).

In the reports of the 10th Census. Describes the landforms and the soils of the state.

Smith, Eugene A., 1884b, Report on the cotton production of the state of Alabama with a discussion of the general agricultural features of the state, HMD 42, part 6, v. 13, part 6, 47-2, p. 1-163, 5 plates and maps, (2134).

In the reports of the 10th Census. Describes the landforms and the soils of the state.

Smith, Eugene A., 1884c, Report on the cotton production of the state of Florida with a discussion of the general agricultural features of the state, HMD 42, part 6, v. 13, part 6, 47-2, p. 1-77, 3 maps, (2134).

In the reports of the 10th Census. Describes the landforms and the soils of the state.

Smith, Eugene A., 1903, The Portland cement materials of central and southern Alabama, SD 19, v. 2, 58-1, p. 12-23, (4563).

Describes the limestone deposits of the area. See also Eckel, Edwin C., 1903.

Smith, J. Lawrence, 1855, Minerals and mineral waters of Chile, HED 121, v. 15, part 2, 33-1, p. 85-107, (729).

Describes and gives chemical analyses of the minerals and mineral waters. In v. 2 of the U. S. Astronomical Expedition to the Southern Hemisphere. See also Gilliss, J. M., 1855.

Smith, J. Lawrence, 1856, Lecture on meteoric stones, SMD 73, v. 1, 34-1, p. 151-174, (835).

Also in HMD 113, v. 2, 34-1, p. 151-174, (835). A detailed report that includes chemical analyses of meteorites. In the Annual Report of the Smithsonian Institution.

Smith, Warren D., 1905a, Advance report to the chief of the mining bureau upon the coal deposits of Batan Island, HD 2, v. 14, 59-1, p. 679-701, (4954).

Describes the coal deposits of this part of the Philippines.

Smith, Warren D., 1905b, Narrative report of work in Batan Island, HD 2, v. 12, 59-1, p. 312-314, (4952).

Describes the geology and coal deposits of part of the Philippines.

Smith, William G., 1901, Soil survey from Raleigh to Newbern, North Carolina, HD 526, v. 107, 56-2, p. 187-205, 8 plates, 4 maps, (4181).

Smith, William G., 1903a, Soil survey of the Columbus area, Ohio, HD 473, v. 106, 57-2, p. 403-423, 1 plate, (4545).

Smith, William G., 1903b, Soil survey of the Toledo area, Ohio, HD 473, v. 106, 57-2, p. 383-402, 3 plates, 1 map, (4545).

Smith, William G., 1904, Soil survey of the Viroqua area, Wisconsin, HD 746, v. 111, 58-2, p. 799-814, 1 map (in v. 112), (4737-4738).

Smith, William G., and **Bennett, Frank, Jr.,** 1902, Soil survey of the Lebanon area, Pennsylvania, HD 655, v. 117, 57-1, p. 149-171, 4 plates, 1 map, (4384).

Smith, William G., and **Bennett, Hugh H.,** 1904, Soil survey of Davidson County, Tennessee, HD 746, v. 111, 58-2, p. 605-617, 1 map (in v. 112), (4737-4738).

Smith, William G., and **Carter, William T., Jr.,** 1903, Soil survey of the Smedes area, Mississippi, HD 473, v. 106, 57-2, p. 325-348, 3 plates, 2 maps, (4545).

Smith, William G., and **Carter, William T., Jr.,** 1904a, Soil survey of the Fort Valley area, Georgia, HD 746, v. 111, 58-2, p. 317-330, 1 map (in v. 112), (4737-4738).

Smith, William G., and **Carter, William T., Jr.,** 1904b, Soil survey of the McNeill area, Mississippi, HD 746, v. 111, 58-2, p. 405-418, 1 map (in v. 112), (4737-4738).

Smith, William G., and **Coffey, George N.,** 1904, Soil survey of the Craven area, North Carolina, HD 746, v. 111, 58-2, p. 253-278, 1 map (in v. 112), (4737-4738).

Smith, William G., and **Martin, J. O.,** 1902, Soil survey of Hartford County, Maryland, HD 655, v. 117, 57-1, p. 211-237, 1 map, (4384).

Smith, William G., and **Meeker, F. N.,** 1904, Soil survey of Sumter County, Alabama, HD 458, v. 108, 58-3, p. 317-342, 1 map, (4887-4888).

Smith, William G., and **Meeker, F. N.,** 1907, Soil survey of Blount County, Alabama, HD 925, v. 110, 59-1, p. 407-424, 1 map (in v. 111), (5050-5051).

Sollas, W. J., 1899, Funafuti: the story of a coral atoll, HD 309, v. 91, 55-3, p. 389-406, (3833).

Describes reefs and discusses their origin. In the Annual Report of the Smithsonial Institution.

Sollas, W. J., 1901, Evolutional geology, HD 537, part 1, v. 114, 56-2, p. 289-314, 1 map, (4188).

Discusses the causes of orogeny and the general evolution of the earth. In the Annual Report of the Smithsonian Institution.

Soule, Frank, 1902, Irrigation from the San Joaquin River, SD 356, v. 27, 57-1, p. 215-258, 3 plates, (4246).

Describes the fluvial and drainage features of this part of California. See also Manson, Marsden, 1902; and Nutter, Edward H., 1902.

Spear, John C., 1872, Report on the geology, mineralogy, natural history, inhabitants, and agriculture of the Isthmus of Tehuantepec, SED 6, v. 3, 42-2, p. 101-139, (1480).

See also Schufeldt, Robert W., 1872.

Spurr, J. E., and **Post, W. S.,** 1899, Report of the Kuskokwim expedition, SD 172, v. 13, 55-3, p. 28-39, (3737).

Describes the geology and economic deposits of this part of Alaska.

Stansbury, Howard, 1852, Exploration and survey of the valley of the Great Salt Lake of Utah including a reconnaissance of a new route through the Rocky Mountains, SED 3, v. 2, 32-special, p. 1-487, 54 plates, 2 maps, (608).

Includes descriptions of the landforms, lithology, and fossils. The 4 paleontological plates are not found in all copies of this Senate edition. Some copies of this Senate edition bear the date 1853.

Stebbins, Rufus P., 1874, Discourse of Louis Agassiz, SMD 130, v. 2, 43-1, p. 198-210, (1585).

Describes the life and the work of Agassiz. In the Annual Report of the Smithsonian Institution.

Stefansson, Jon, 1907, Iceland: its history and inhabitants, HD 575, v. 97, 59-2, p. 275-294, (5200).

Describes the earthquakes, volcanoes, and general geology of the island. In the Annual Report of the Smithsonian Institution.

Steinberger, A. B., 1874, Report upon Samoa or the Navigator's Islands, SED 45, v. 2, 43-1, p. 1-58, (1581).

Includes brief descriptions of the topography and other geologic features.

Stevens, I. I., 1855a, Report of explorations for a route for the Pacific Railroad near the 47th and 49th parallels of north latitude, from St. Paul to Puget Sound, SED 78, v. 13, part 1, 33-2, p. 1-636, (758).

Also in HED 91, v. 11, part 1, 33-2, p. 1-636, (791); and also in HED 129, v. 18, part 1, 33-1, p. 1-599, (736). Part of the Pacific Railroad Survey Reports. This describes the exploration of the northernmost route.

Stevens, I. I., 1855b, Supplementary report of explorations for a route near the 47th and 49th parallels . . . , from St. Paul to Puget Sound, SED 40, v. 18, 35-2, p. 1-333, 70 plates, 3 maps, (992).

Also in HED 56, v. 11, part 1, 36-1, p. 1-358, 70 plates, 3 maps, (1054). Although publication date is printed as 1855 this document actually came out in 1860. A supplemental volume of the Pacific Railroad Survey Reports. This describes the northern route and its geology in some detail.

Stevenson, J. T., 1879, Preliminary report(s) of a special geological party operating in Colorado and New Mexico, from Spanish Peaks to the south . . . , HED 1, part 2, v. 5, 46-2, p. 2249-2259, (1906).

Describes the geology of parts of Colorado and New Mexico. In Wheeler, George M., 1879a.

Stockton, John C., et al, 1845, (Reports on the mineral lands of Lake Superior), SED 175, v. 11, 28-2, p. 1-22, 1 map, (461).

Describes the topography and mineral deposits of part of the Lake Superior region.

Suess, Edward, 1873, The boundary line between geology and history, HMD 107, v. 3, 42-3, p. 223-232, (1573).

Describes the Pleistocene geology of Europe and the first appearance of man in Europe. In the Annual Report of the Smithsonian Institution.

Tarr, R. S., 1905, Gorges and waterfalls of central New York, HD 460, v. 111, 58-3, p. 136-137, (4890).

An abstract in the Report of the 8th International Geographic Congress.

Tassin, Wirt, 1897, The mineralogical collection in the U. S. National Museum, HD 425, part 2, v. 82, 54-1, p. 995-1000, 1 plate, (3449).

In the Annual Report of the U. S. National Museum.

Tassin, Wirt, 1899a, Catalogue of the series illustrating the properties of minerals, HD 575, part 2, v. 79, part 1, 55-2, p. 649-688, (3707).

In the Annual Report of the U. S. National Museum.

Tassin, Wirt, 1899b, Classification of the mineral collection in the U. S. National Museum, HD 575, part 2, v. 79, part 1, 55-2, p. 747-810, (3707).

In the Annual Report of the U. S. National Museum.

Tassin, Wirt, 1902a, Descriptive catalogue of the collections of gems in the United States National Museum, HD 537, part 2, v. 115, 56-2, p. 473-670, 9 plates, (4189).

In the Annual Report of the U. S. National Museum.

Tassin, Wirt, 1902b, Descriptive catalogue of the meteorite collection in the United States National Museum, to January 1, 1902, HD 537, part 2, v. 115, 56-2, p. 671-698, 4 plates, (4189).

In the Annual Report of the U. S. National Museum.

Taylor, Bayard, 1856, Report of an exploration of Peel Island, HED 97, v. 12, part 2, 33-2, p. 67-71, (803).

Also in SED 79, v. 14, part 2, 33-2, p. 67-71, (770). Describes the landforms and general geology. In Perry, M. C., 1856.

Taylor, James W., 1867, Report . . . upon gold and silver mining east of the Rocky Mountains, HED 92, v. 11, 39-2, p. 2-28, (1293).

An extensive report on the known deposits. This report along with Browne, J. R., 1867 forms the start of a series which eventually becomes "The mineral resources of the United States" published by the U. S. Geological Survey.

Taylor, James W., 1868, Report . . . upon gold and silver mining east of the Rocky Mountains, HED 273, v. 17, 40-2, p. 1-71, (1343).

Describes the mines and economic geology of these metals. See also Taylor, James W., 1867.

Taylor, F. W., 1903, Soil survey of the Abbeville area, South Carolina, HD 473, v. 106, 57-2, p. 273-289, 1 map, (4545).

Teall, J. J. Harris, 1903, The evolution of petrological ideas, HD 484, part 1, v. 109, 57-2, p. 287-308, (4548).

Describes the history of petrology. In the Annual Report of the Smithsonian Institution.

Tharp, W. E., and **Mann, Charles J.,** 1908, Soil survey of Greene County, Indiana, HD 352, v. 75, 59-2, p. 755-789, 1 map (in v. 76), (5178-5179).

Thomas, Cyrus, 1872, Agricultural resources of the territories, HED 326, v. 15, 42-2, p. 207-279, (1520).

Describes the landforms of parts of the Rocky Mountains, Great Plains, and Great Basin. In Hayden, F. V., 1872a.

Thomas, Cyrus, 1873, Physical geography and agricultural resources of Minnesota, Dakota, and Nebraska, HMD 112, v. 3, 42-3, p. 275-313, (1573).

Describes the landforms of the areas. In Hayden, F. V., 1873.

Thomson, J. P., 1898, The physical geography of Australia, HD 352, part 1, v. 72, 54-2, p. 245-272, (3548).

Describes the landforms and major features of the continent. In the Annual Report of the Smithsonian Institution.

Thoulet, J., 1905, Carte bathymétrique générale des océans, HD 460, v. 111, 58-3, p. 439-444, (4890).

Describes the known features of the bottom of the oceans. In the Report of the 8th International Geographic Congress.

Trowbridge, W. P., 1860, Description of an apparatus . . . and of the method of applying it in determining ocean depths and obtaining specimens of bottom, HED 41, v. 7, 36-1, p. 359-361, 1 plate, (1049).

In the Annual Report of the Coast Survey.

True, Frederick W., 1898, An account of the United States National Museum, HD 352, part 2, v. 73, 54-2, p. 287-324, (3549).

Describes the formation and history of the museum. In the Annual Report of the U. S. National Museum.

Tyson, P. T., 1850, Report on the geology of California, SED 47, part 1, v. 10, 31-1, p. 3-74, 10 plates and maps, (558).

Describes the geology of the Sierra Nevada, the Coast Ranges, the Sacramento Valley, and the economic geology of these areas.

U. S. Army Corps of Engineers, 1874, Report on improvement of the mouths of the Mississippi, HED 1, part 2, v. 3, 43-2, p. 776-888, 6 maps, (1636).

Describes the fluvial and shore features of portions of the Mississippi Delta.

U. S. Army Corps of Engineers, 1903, Analytical and topical index to the reports of the Chief of Engineers and officers of the Corps of Engineers, United States Army, 1866-1900, HD 439, v. 93-95, 57-2, p. 1-1788, (4532-4534).

A complete 3 volume index. Many of the included reports contain material of geological nature.

U. S. Army Corps of Engineers, 1905, Defenses of the Cape Fear River, North Carolina, HD 2, v. 8, 59-1, p. 3010-3015, 3 plates, (4948).

Describes the shore features of this part of the North Carolina coast.

U. S. Department of Agriculture, 1869a, Concentrated fertilizers in the southern states, HED—no document number, v. (15), 40-3, p. 396-404, (1383).

Describes the calcareous marls, greensands, and bone beds of the region.

U. S. Department of Agriculture, 1869b, The Marl region of Virginia, HED—no document number, v. (15), 40-3, p. 389-395, (1383).

Describes the Tertiary beds used for agricultural fertilizer.

U. S. Department of Agriculture, 1869c, Mineral fertilizers of the Atlantic states, HED—no document number, v. (15), p. 367-388, (1383).

Describes the lithology and extent of the strata utilized as fertilizer.

U. S. Department of Agriculture, 1870, Mineral fertilizers of the Mississippi Valley, HED—no document number, v. 14, 40-2, p. 548-583, 3 maps, (1428).

Describes the Tertiary sedimentary rocks used as fertilizers.

U. S. Department of Agriculture, 1871, The Great Salt Lake Basin, HED—no document number, v. (13), 41-3, p. 559-569, (1461).

Includes a brief description of the landforms of the area.

U. S. Department of Agriculture, 1872, Agricultural topography and resources of Montana Territory, HED 327, v. 17, 42-2, p. 431-448, 1 plate, (1522).

Describes the landforms and mineral resources of the area.

U. S. Department of Agriculture, 1890, A report on the preliminary investigation to determine the proper location of Artesian wells within the area of the 97th meridian and east of the foothills of the Rocky Mountains, SED 222, v. 12, 51-1, p. 5-398, (2689).

Describes the geology and hydrology of the area and compares it with similar areas in Algeria and Australia.

U. S. Department of Commerce and Labor, 1905, Formation of coal and petroleum, HD 334, v. 61, 58-3, p. 196-198, (4840).

Brief general description of the origin of coal and petroleum.

U. S. Department of the Interior, 1866, Map of the United States and Territories showing deposits of economic resources, HED 1, v. 2, 39-2, between p. 496-497, 1 map, (1284).

A map without accompanying text.

U. S. War Department, 1855a, Maps to illustrate the various reports of surveys . . . of the Pacific Railroad Reports, HED 129, v. 18, part 4, 33-1, 13 maps, (739).

See also U. S. War Department, 1855b.

U. S. War Department, 1855b, Topographical maps, profiles, and sketches, to illustrate the various reports of surveys . . . of the Pacific Railroad Reports, SED 78, v. 13, part 11, 33-2, 23 plates, 23 maps, (768).

Also in HED 91, v. 11, part 11, 33-2, 23 plates, 23 maps, (801). See also U. S. War Department 1855a.

U. S. War Department, 1874, Report . . . relative to geographical and geological surveys west of the Mississippi, HED 240, v. 16, 43-1, p. 1-15, 1 map, (1614).

A report, including letters by Powell and Hayden, dealing with the duplication of work by the various surveys.

U. S. War Department, 1878, Views of the War Department concerning the public surveys of the territories of the United States, HED 1, part 2, v. 5, 45-3, p. 1661-1666, (1846).

Presents the case for keeping the surveys under the direction of the War Department. See also Humphreys, A. A., 1878a.

U. S. War Department, 1905, Extract from the report of the Secretary of War for the year 1905 (on Batán coal fields), HD 2, v. 14, 59-1, p. 677-678, (4954).

Describes the coal deposits of part of the Philippines.

U. S. War Department, 1906a, The Isle of Pines: its situation, physical features, inhabitants, resources, and industries, SD 311, v. 6, 59-1, p. 1-43, 2 maps, (4914).

Describes the landforms and economic resources of this West Indian Island.

U. S. War Department, 1906b, (Miscellaneous correspondence relating to the Isle of Pines), SD 205, v. 5, 59-1, p. 23-277, (4913).

Includes descriptions of some of the landforms and economic deposits on this West Indian island.

Van Hise, C. R., 1893, Excursion to Lake Superior; Pre-Cambrian geology of the Lake Superior region, HMD 107, v. 13, 53-2, p. 489-512, (3241).

A field guide for a trip conducted by the 5th International Geologic Congress. In Congrès Geologique International, 1893.

Van Ingen, H. S., 1874, Special report on improvement of Edgartown Harbor, Massachusetts, HED 1, part 2, v. 4, 43-2, p. 183-216, (1637).

Describes the shore features and the historical changes in these features.

Van Lennep, 1868, Geological report for 1867 on the Lone Tree Valley and the country west of the Black Hills, on the line of the Union Pacific railroad company, HED 331, v. 20, 40-2, p. 48-71, (1346).

Includes cross-sections, lithologic descriptions, and descriptions of the coal beds of the area. See also Dodge, G. M., 1868.

Van Vleet, A. H., 1904, Geology and Natural History (of Oklahoma), HD 5, v. 22, 58-3, p. 474-478, 1 map, (4801).

Includes a description of the gypsum and salt deposits.

Van Vleet, A. H., 1907, Geology and Natural History (of Oklahoma), HD 5, v. 16, 59-2, p. 203-208, (5119).

Describes the geology and mineral resources.

Wade, B. F., White, A. D., and **Howe, S. G.,** 1871, Report on the Island of Santo Domingo, SED 9, v. 1, 42-1, p. 4-34, (1466).

Includes a brief description of the areal geology.

Walcott, Charles D., 1894, Geologic time, as indicated by the sedimentary rocks on North America, HMD 184, part 1, v. 29, 53-2, p. 301-334, 1 map, (3257).

Discusses the measurement of geologic time by the rates of deposition of sedimentary rocks. In the Annual Report of the Smithsonian Institution.

Walcott, Charles D., 1897, Progress and result of the Geological Survey exploration of Alaska during the season of 1896, SD 109, v. 4, 54-2, p. 1-6, 1 map, (3470).

Includes a description of the known gold deposits.

Walcott, Charles D., 1901, Geology, etc., of the Coosa Valley, Alabama, SD 65, v. 5, 56-2, p. 1-4, (4033).

Describes the geology and economic deposits of this part of the state. See also Walcott, Charles D., 1905.

Walcott, Charles D., 1902, Abstract of reports, letters, clippings, indorsements, resolutions, petitions, and requests showing the demand for the topographical work of the United States Geological Survey, SD 136, v. 13, 57-1, p. 4-67, (4231).

Walcott, Charles D., 1905, (The geology and natural resources of the Coosa Valley, Alabama), HD 219, v. 52, 58-3, p. 29-30, (4831).

See also Walcott, Charles D., 1901.

Walcott, Charles D., 1906, Black sands of the Pacific slope, SD 65, v. 2, 59-1, p. 1-24, (4910).

Describes heavy mineral sands and their placer deposits along the western part of the continent.

Waldeyer, Charles, 1873, Hydraulic mining in California, HED 210, v. 9, 42-3, p. 390-424, 17 plates, (1567).

A part of Raymond, Rossiter W., 1873.

Wall, G. P., and **Sawkins, Joseph,** 1857, Report of progress, from August 25, 1856, to February 24, 1857, of the survey of the economic geology of Trinidad, SMD 54, v. 1, 34-3, p. 281-288, (890).

Describes the geology and mineral resources. In the Annual Report of the Smithsonian Institution.

Wallace, A. R., 1894, The ice age and its work, HMD 184, part 1, v. 29, 53-2, p. 277-300, (3257).

Describes erosion and deposition by Pleistocene glaciers. In the Annual Report of the Smithsonian Institution.

Ward, Lester F., 1901, The petrified forests of Arizona, HD 737, part 1, v. 119, 56-1, p. 289-307, 3 plates, (4016).

Describes the scenery, geology, and petrified trees. In the Annual Report of the Smithsonian Institution.

Ward, Willard P., 1869, The Prussian Royal School of Mines at Berlin, HED 54, v. 9, 40-3, p. 244-249, (1374).

In Raymond, Rossiter W., 1869.

Warren, G. K., 1856, Explorations in the Dacota Country, in the year 1855, SED 76, v. 13, 34-1, p. 2-79, 2 maps, (822).

Describes the landforms and geology of parts of the Dakotas, Wyoming, and Nebraska.

Warren, G. K., 1858, Preliminary report on explorations in Nebraska, SED 1, v. 2, 35-2, p. 62-670, (975).

Describes the exploration and geology of parts of Nebraska, South Dakota, Kansas, Colorado, and Wyoming.

Warren, G. K., 1875a, Report and plan for the reclamation of the alluvial basin of the Mississippi River subject to inundation, HED 127, v. 15, 43-2, p. 5-160, 1 plate, 4 maps, (1648).

Also in HED 1, part 2, v. 4, 44-1, p. 536-678, 1 plate, (1675). Describes the topographic and fluvial features of the lower Mississippi River. See also Forshey, C. G., 1875.

Warren, G. K., 1875b, Report of the result of the examination and survey of the Minnesota River, HED 76, v. 12, 43-2, p. 2-72, 5 maps, (1645).

Also in HED 1, part 2, v. 4, 44-1, p. 380-451, (1675). Describes the fluvial and topographic features and includes a history of the surveys of the river.

Warren, G. K., 1876, Report on the transportation route along the Wisconsin and Fox Rivers . . . between the Mississippi River and Lake Michigan, SED 28, v. 1, 44-1, p. 1-114, 8 plates, 11 maps, (1664).

Also in HED 1, part 2, v. 4, 44-2, p. 189-298, 8 plates, 11 maps, (1744). Describes the fluvial and topographic features of this part of Wisconsin.

Warren, G. K., 1878, Report on bridging the Mississippi between Saint Paul, Minn., and Saint Louis, Mo., SED 69, v. 5, 45-2, p. 7-232, 29 maps and diagrams, (1784).

Describes the fluvial features of the area.

Weed, Walter H., 1893, Geysers, HMD 334, part 1, v. 43, 52-1, p. 163-178, (3001).

Describes geysers in Wyoming, New Zealand, and Iceland and discusses the origin of geysers. In the Annual Report of the Smithsonian Institution.

Weed, Walter H., 1902, Geological sketch of the Hot Springs district, Arkansas, SD 282, v. 20, 57-1, p. 79-94, 7 plates, 3 maps, (4239).

Describes the areal geology and the hot springs. See also Haywood, J. K., 1902.

Wheeler, E. S., 1902, Earthquakes (in Central America), SD 357, v. 26, 57-1, p. 63-68, (4245).

Describes the earthquakes and their possible effects on an interoceanic canal. See also Dutton, C. E., 1902.

Wheeler, George M., 1872a, Preliminary report of explorations in Nevada and Arizona, SED 65, v. 2, 42-2, p. 2-94, 1 map, (1479).

See also Barlow, J. W., 1872; and also Gilbert, G. K., 1872; and also Lockwood, Daniel W., 1872; and also Lyle, D. A., 1872.

Wheeler, George M., 1872b, Report . . . of explorations and surveys in Nevada, Utah, and Arizona, HED 1, part 2, v. 8, 42-3, p. 1124-1126, (1559).

A progress report on the work under Wheeler's direction.

Wheeler, George M., 1873, Annual report . . . upon explorations and surveys west of the 100th meridian, in Nevada, Utah, Colorado, New Mexico, and Arizona, HED 1, part 2, v. 2, 43-1, p. 1211-1218, 1 map, (1598).

A progress report on the work under Wheeler's direction.

Wheeler, George M., 1874, Annual report . . . for the fiscal year ending June 30, 1874, HED 1, part 2, v. 4, 43-2, p. 480-588, 1 map, (1637).

A progress report on the work under Wheeler's direction. See also Cope, E. D., 1874.

Wheeler, George M., 1875, Annual report . . . for the fiscal year ending June 30, 1875, HED 1, part 2, v. 5, 44-1, p. 921-1108, 7 plates, 2 maps, (1676).

A progress report on the work under Wheeler's direction. See also Cope, E. D., 1875; and also Loew, Oscar, 1875.

Wheeler, George M., 1876, Annual report . . . for the year ending June 30, 1876, HED 1, part 2, v. 5, 44-2, p. 219-563, 1 plate, (1745).

A progress report on the work under Wheeler's direction. See also Conkling, A. R., 1876; and also Loew, Oscar, 1876a and 1876b; and also Marcou, Jules, 1876.

Wheeler, George M., 1877, Annual report . . . for the fiscal year ending June 30, 1877, HED 1, part 2, v. 4, 45-2, p. 1209-1334, 1 plate, 2 maps, (1796).

A progress report on the work under Wheeler's direction. See also Church, John A., 1877 and also Conkling, A. R., 1877a, 1877b, and 1877c.

Wheeler, George M., 1878, Annual report . . . for the fiscal year ending June 30, 1878, HED 1, part 2, v. 5, 45-3, p. 1429-1651, 5 plates, 2 maps, (1846).

A progress report on the work under Wheeler's direction. See also Church, John A., 1878; and also Conkling, A. R., 1878.

Wheeler, George M., 1879a, Annual report for the fiscal year ending June 30, 1879, HED 1, part 2, v. 5, 46-2, p. 1977-2313, 5 maps, (1906).

A progress report on the work under Wheeler's direction. See also Stevenson, J. T., 1879.

Wheeler, George M., 1879b, Letter . . . in regard to the total amount expended in the prosecution of the surveys west of the 100th meridian, HED 104, v. 16, 45-3, p. 1-2, (1858).

Describes and lists the main expenses of Wheeler's surveys.

Wheeler, George M., 1880, Report . . . for the fiscal year ending June 30, 1880, HED 1, part 2, v. 5, 46-3, p. 2459-2499, 1 map, (1955).

A progress report on the work under Wheeler's direction.

Wheeler, George M., 1884, Report . . . for the fiscal year ending June 30, 1884, HED 1, part 2, v. 5, 48-2, p. 2375-2378, (2280).

A progress report on the work under Wheeler's direction.

Wheeler, George M., 1885, Report upon the third International Geographic Congress and Exhibition at Venice, Italy, 1881, HED 270, v. 18, 48-2, p. 1-586, 11 maps, (2293).

Describes the various national land and marine surveys of the world.

Wheelock, T. B., 1834, (Journal of the campaign of the regiment of dragoons for the summer of 1834), SED 1, v. 1, 23-2, p. 73-93, (266).

Also in SED 209, 24-1, (281), but this document is missing from nearly all copies of the Documents Set. Includes landform and lithologic description and includes some stratigraphic information. Is a report on the expedition across the Great Plains to the Rocky Mountains under Col. Henry Dodge.

Whipple, A. W., 1851, Extract from a journal of an expedition from San Diego, California to the Rio Colorado, from September 11, to December 11, 1849, SED 19, v. 3, 31-2, p. 2-19, (589).

Describes the landforms and some of the lithology.

Whipple, A. W., 1855, Report of explorations for a railway route near the 35th parallel of north latitude, from the Mississippi River to the Pacific Ocean, SED 78, v. 13, part 3, 33-2, p. 1-136 and p. 1-77, 20 plates, (760).

Also in HED 91, v. 11, part 3, 33-2, p. 1-136 and p. 1-77, 20 plates (793); and also in HED 129, v. 18, part 2, 33-1, p. 1-154, (739). Part of the Pacific Railroad Survey Reports.

White, Charles A., 1883, Contributions to invertebrate paleontology, HMD 62, part 1, v. 21, 47-1, p. 1-171, 32 plates, (2056).

Describes the invertebrate fossils of parts of Wyoming, Idaho, Utah, Colorado, Iowa, Illinois, Missouri, and Indiana. In Hayden, F. V., 1883a.

White, Charles A., 1893, The relation of biology to geological investigation, HMD 114, part 2, v. 23, 52-2, p. 245-368, (3132).

Discusses paleontological dating and correlation. In the Annual Report of the U. S. National Museum.

White, David, 1903, The American range of the Cycadofilices, HD 460, v. 111, 58-3, p. 616, (4890).

An abstract in the Report of the 8th International Geographic Congress.

Whitfield, Benjamin, 1874, Report of the geologist of the Nicaraguan exploring expedition, SED 57, v. 3, 43-1, p. 25-30, (1582).

Describes the geology of part of the country. In Hatfield, Chester, 1874.

Whitfield, R. P., 1876, Descriptions of new fossils (collected on a reconnaissance from Carroll, Montana, to the Yellowstone National Park), HED 1, part 2, v. 5, 44-2, p. 694-699, 2 plates, (1745).

Describes the Cretaceous invertebrates from Montana. In Ludlow, William, 1876.

Whiting, Henry L., 1875, Report on shore-line changes at Edgartown Harbor, Massachusetts, HED 240, v. 12, 42-3, p. 262-265, (1570).

In the Annual Report of the Coast Survey.

Whiting, Henry L., 1887, Reports of changes in the shore-line and beaches of Martha's Vineyard, as derived from comparisons of recent with former surveys, HED 40, v. 22, 49-2, p. 263-266, 1 map, (2481).

Describes the changes in coastal features for this part of Massachusetts. In the Annual Report of the Coast and Geodetic Survey.

Whiting, Henry L., 1890, Recent changes in the south inlet into Edgartown Harbor, Martha's Vineyard, HED 55, v. 27, 51-1, p. 459-460, (2742).

Describes the shoreline changes in this part of Massachusetts. In the Annual Report of the Coast and Geodetic Survey.

Whiting, William H. C., 1850, Report . . . on the reconnaissance of the western frontier of Texas, SED 64, v. 14, 31-1, p. 236-250, (562).

Describes the landforms and rocks of parts of west Texas.

Whitney, J. D., 1848a, (Report on the geology of) the northern peninsula of Michigan, SED 2, v. 2, 30-1, p. 223-230, (504).

A preliminary report on Whitney's work for the summer of 1848. In Jackson, Charles T., 1848a.

Whitney, J. D., 1848b, (Synopsis of a report on the geology of part of the Lake Superior region), SED 2, v. 2, 30-2, p. 154-159, (530).

In Jackson, Charles T., 1848b.

Whitney, J. D., 1849, Field notes for 1847, SED 1, v. 3, 31-1, p. 713-758, (551).

Also in HED 5, v. 3, part 3, 31-1, p. 713-758, (571). A part of Jackson's final report. See Jackson, Charles T., 1849.

Whitney, J. D., 1874, Physical features of the United States, HMD— unnumbered, v. 3, 42-1, p. 1-4, (1473).

In the reports of the 9th Census.

Wigmore, H. L., 1904, Coal mining in the Philippine Islands, HD 2, v. 9, 58-3, p. 3870-3875, (4788).

Describes the coal resources and mining.

Wigmore, H. L., 1905a, Report of an examination of the coal deposits of Polillo Island, HD 2, v. 14, 59-1, p. 702-706, (4954).

Describes the coal deposits of this part of the Philippines.

Wigmore, H. L., 1905b, Report of examination of coal deposits on the Batán Military Reservation, Batán Island, HD 2, v. 14, 59-1, p. 653-676, 7 maps, (4954).

Describes the coal deposits of this part of the Philippines.

Wilder, Henry J., 1904, Soil survey of the Marshall area, Minnesota, HD 746, v. 111, 58-2, p. 815-831, 1 map (in v. 112), (4737-4738).

Wilder, Henry J., and **Belden, H. L.,** 1905a, Soil survey of Adams County Pennsylvania, HD 458, v. 108, 58-3, p. 119-150, 1 map (in v. 109), (4887-4888).

Wilder, Henry J., and **Belden, H. L.,** 1905b, Soil survey of the Vergennes area, Vermont-New York, HD 458, v. 108, 58-3, p. 73-98, 1 map (in v. 109), (4887-4888).

Wilder, Henry J., and **Bennett, Hugh H.,** 1905, Soil survey of Macon County, Alabama, HD 458, v. 108, 58-3, p. 291-315, 1 map (in v. 109), (4887-4888).

Wilder, Henry J., and **Geib, W. J.,** 1904a, Soil survey of the Pikeville area, Tennessee, HD 746, v. 111, 58-2, p. 577-603, 1 map (in v. 112), (4737-4738).

Wilder, Henry J., and **Geib, W. J.,** 1904b, Soil survey of the Pontiac area, Michigan, HD 746, v. 111, 58-2, p. 659-685, 1 map (in v. 112), (4737-4738).

Wilder, Henry J., and **Shaw, Charles F.,** 1908, Soil survey of the Fayetteville area, Arkansas, HD 352, v. 75, 59-2, p. 587-627, 1 map (in v. 76), (5178-5179).

Wilder, Henry J., Strahorn, A. T., and **Geib, W. J.,** 1907, Soil survey of Montgomery County, Pennsylvania, HD 925, v. 110, 59-1, p. 97-133, 1 map (in v. 111), (5050-5051).

Wilder, Henry J., et al, 1907a, Soil survey of Chester County, Pennsylvania, HD 925, v. 110, 59-1, p. 135-174, 1 map (in v. 111), (5050-5051).

Wilder, Henry J., et al, 1907b, Soil survey of Leon County, Florida, HD 925, v. 110, 59-1, p. 363-388, 1 map (in v. 111), (5050-5051).

Williams, Gardner F., 1906, The genesis of the diamond, HD 930, v. 113, 59-1, p. 193-209, (5053).

In the Annual Report of the Smithsonian Institution.

Williams, Harry G., 1870, (Report on the Stikine River region, Alaska), SED 68, v. 2, 41-2, p. 11-14, (1406).

Includes a description of the glacier and of the mineral resources.

Williams, Henry S., 1890, North American paleontology for 1887 and 1888, HMD 142, part 1, v. 14, 50-2, p. 261-326, (2668).

Includes a bibliography for 1887 and 1888. In the Annual Report of the Smithsonian Institution.

Williamson, R. S., 1850, (Report on limestone collected at Mount Diablo), SED 47, part 2, v. 10, 31-1, p. 34-35, (558).

Briefly describes this part of California and the limestone found there.

Williamson, R. S., 1855, Report of explorations in California to connect with routes near the 35th and 32nd parallels . . . , SED 78, v. 13, part 5, 33-2, p. 1-370, 54 plates, 4 maps, (762).

Also in HED 91, v. 11, part 5, 33-2, p. 1-370, 54 plates, 4 maps, (795). Part of the Pacific Railroad Survey Reports.

Wilkes, Charles, Hunt, H., and **Martin, D. B.,** 1859, Reports on the examination of the iron and coal of Deep River country, in the state of North Carolina, SED 26, v. 7, 35-2, p. 2-27, (981).

Describes the stratigraphy, structure and mineral deposits of this part of the state.

Wilsing, J., 1890, Determination of the mean density of the earth by means of a pendulum principle, HMD 142, part 1, v. 14, 50-2, p. 635-646, (2668).

In the Annual Report of the Smithsonian Institution.

Wilson, A. W. G., 1905, Physiography of the Archean areas of Canada, HD 460, v. 111, 58-3, p. 116-135, 3 plates, 2 maps, (4890).

Describes landforms in parts of Labrador, Quebec, Ontario, Manitoba, Keewatin, and Mackenzie. In the Report of the 8th International Geographic Congress.

Wilson, Joseph S., 1868, Report . . . on the extension of the geological survey to different localities of the public domain, HMD 136, v. 2, 40-2, p. 1-4, (1350).

A letter from the Commissioner of the General Land Office urging geological surveys of the western territories.

Winchell, N. H., 1874, Preliminary geological report . . . (on the Black Hills), HED 1, part 2, v. 4, 43-2, p. 630-632, (1637).

A brief report on the exploration and geology of the area. See also Grinnell, George B., 1874.

Winchell, N. H., 1875, Geological report (on the Black Hills), HED 1, part 2, v. 5, 44-1, p. 1131-1177, 1 map, (1676).

In Ludlow, William, 1875.

Wislizenus, A., 1848, Memoir of a tour to northern Mexico . . . in 1846 and 1847, SMD 26, v. 1, 30-1, p. 1-141, 1 plate, 2 maps, (511).

Describes the landforms and some of the rocks and fossils and includes a "geological sketch".

Woodward, Robert S., 1891, The mathematical theories of the earth, HMD 129, part 1, v. 10, 51-2, p. 183-200, (2878).

Describes various theories about the origin of the earth. In the Annual Report of the Smithsonian Institution.

Workman, Fanny B., 1905, First exploration of Hoh Lumba and Sosbon glaciers: two record ascents in the Himalayas, HD 460, v. 111, 58-3, p. 724-731, 1 plate, (4890).

Describes the glacial features of parts of Kashmir. In the Report of the 8th International Geographic Congress.

Workman, W. H., 1905, The moraines of the Chogo Lungma glacier in Baltistan, HD 460, v. 111, 58-3, p. 732-736, 1 plate, (4890).

Describes the glacial features of this part of Kashmir. In the Report of the 8th International Geographic Congress.

Wright, H. G., 1879, Report . . . upon a provision in the sundry civil bill, approved March 3, 1879, discontinuing the geographical surveys west of the 100th meridian . . . , SED 11, v. 1, 46-1, p. 1-3, (1869).

Letter from the acting Chief of Engineers requesting funds to continue the War Department's surveys of the territories.

Wright, H. G., 1880, . . . the importance of the geographical and topographical surveys of the territories of the United States west of the Mississippi River . . . , SED 118, v. 4, 46-2, p. 1-2, (1885).

See previous annotation.

Wright, H. G., Foster, J. G., and **Newcomb, Wesley,** 1872, Report of the Sutro Tunnel Commission, SED 15, v. 1, 42-2, p. 3-22, (1478).

Describes the economic deposits of the Comstock Lode in Nevada.

Wyman, Jeffries, 1855, Description of a portion of a lower jaw and a tooth of the *Mastodon andium* . . . , HED 121, v. 15, part 2, 33-1, p. 273-281, 2 plates, (729).

Describes a fossil mastodon found in Chile. In v. 2 of the U. S. Astronomical Expedition to the Southern Hemisphere. See also Gilliss, J. M., 1855.

Younghusband, Frank, 1906, Geographical results of the Tibet mission, HD 930, v. 113, 59-1, p. 265-277, 4 plates, (5053).

Describes the landforms of the area. In the Annual Report of the Smithsonian Institution.

Part 4

INDEX

A

Africa
  Geomorphology
    Geysers
      Peale, A. C., 1883
Age of earth. See **Age** under **Earth.**
Alabama
  Areal geology
    Eckel, Edwin C., & Crider, A. F., 1905
    Smith, Eugene A., 1884b
    Walcott, Charles D., 1901
    Walcott, Charles D., 1905
  Economic geology
    Coal, Iron
      Alabama, University of, 1902
    Iron
      Pumpelly, Raphael, 1886
    Limestone
      Eckel, Edwin C., 1903
      Eckel, Edwin C., & Crider, A. F., 1905
      Smith, Eugene A., 1903
    Mineral resources
      Walcott, Charles D., 1901
      Walcott, Charles D., 1905
  Geomorphology
    Landform description, Soils
      Smith, Eugene A., 1884b

Soils, Blount Co.
  Smith, William G., & Meeker, F. N., 1907
Soils, Dallas Co.
  Carr, E. P., et al, 1907
Soils, Huntsville
  Bennett, Frank, Jr., & Griffen, A. M., 1904
Soils, Lauderdale Co.
  Bonsteel, F. E., et al, 1907
Soils, Lee Co.
  Hearn, W. Edward, & Geib, W. J., 1908
Soils, Macon Co.
  Wilder, Henry J., & Bennett, Hugh H., 1905
Soils, Mobile
  Avon-Burke, R. T., et al, 1904
Soils, Montgomery Co.
  McLendon, W. E., & Mann, Charles J., 1907
Soils, Payne
  Jones, Grove B., & Carr, M. E., 1904
Soils, Perry Co.
  Avon-Burke, R. T., et al, 1903
Soils, Sumter Co.
  Smith, William G., & Meeker, F. N., 1905

Alaska

Areal geology

Allen, Henry T., 1887
Blake, Theodore A., 1869
Brooks, Alfred H., 1899
Brooks, Alfred H., 1905
Brooks, Alfred H.; Richardson, George B., & Collier, Arthur J., 1901
Eldridge, G. H., 1898a
Eldridge, G. H., 1898b
Eldridge, G. H., 1899
Eldridge, G. H., & Muldrow, Robert, 1899
Emmons, S. F., 1898
Mendenhall, Walter C., 1898a
Mendenhall, Walter C., 1898b
Mendenhall, Walter C., 1898c
Mendenhall, Walter C., 1899
Mendenhall, Walter C., 1901
Peters, W. J., & Brooks, Alfred H., 1899
Schrader, Frank C., 1898
Schrader, Frank C., 1899
Schrader, Frank C., & Spencer, Arthur C., 1901
Senate Committee on Military Affairs, 1900
Spurr, J. E., & Post, W. S., 1899

Economic geology

Coal

Barnard, E. C., 1899
Brooks, Alfred H., 1899
Eldridge, G. H., 1898a
Eldridge, G. H., & Muldrow, Robert, 1899
Mendenhall, Walter C., 1899
Peters, W. J., & Brooks, Alfred H., 1899
Schrader, Frank C., 1899

Copper

Peters, W. J., & Brooks, Alfred H., 1899
Schrader, Frank C., 1899
Schrader, Frank C., & Spencer, Arthur C., 1901

Gold

Barnard, E. C., 1899
Brooks, Alfred H., 1899
Brooks, Alfred H.; Richardson, George B., & Collier, Arthur J., 1901
Eldridge, G. H., & Muldrow, Robert, 1899
Emmons, S. F., 1898
Mendenhall, Walter C., 1899
Mendenhall, Walter C., 1901
Peters, W. J., & Brooks, Alfred H., 1899
Schrader, Frank C., 1899
Schrader, Frank C., & Brooks, Alfred H., 1900
Schrader, Frank C., & Spencer, Arthur C., 1901
Walcott, Charles D., 1897

Mineral resources

Abercrombie, W. R., 1900
Blake, Theodore A., 1869
Browne, J. R., 1867
Browne, J. R., 1868
Bryant, Charles, 1870
Bulkley, Charles S., 1868
Davidson, George, 1868
Eldridge, G. H., 1898b
Eldridge, G. H., 1899
McIntyre, H. H., 1870
Mendenhall, Walter C., 1898a
Mendenhall, Walter C., 1898b

Mendenhall, Walter C., 1898c
Morris, William G., 1879
Raymond, Charles W., 1871
Raymond, Rossiter W., 1874
Senate Committee on Military Affairs, 1900
Schrader, Frank C., 1898
Spurr, J. E., & Post, W. S., 1899
Williams, Harry G., 1870

Engineering geology

Railroads
Powell, John W., 1886b

Geological exploration
Allen, Henry T., 1887
Barnard, E. C., 1899
Blake, Theodore A., 1869
Davidson, George, 1868
Eldridge, G. H., & Muldrow, Robert, 1899
Mendenhall, Walter C., 1899
Peters, W. J., & Brooks, Alfred H., 1899
Raymond, Charles W., 1871
Schrader, Frank C., 1899
Senate Committee on Military Affairs, 1900
Spurr, J. E., & Post, W. S., 1899
Walcott, Charles D., 1897

Geomorphology

Landform description
Blake, Theodore A., 1869
Brooks, Alfred H., 1905
Bryant, Charles, 1870
Bulkley, Charles S., 1868
Cook, Frederick A., 1905
Davidson, George, 1868
McIntyre, H. H., 1870
Raymond, Charles W., 1871

Reid, Harry F., 1892
Schwatka, Frederick, 1884

Glacial geology

Muir glacier
Reid, Harry F., 1892

Stikine River glaciers
Williams, Harry G., 1870

Volcanology
Merriam, C. Hart, 1902

Alberta

Engineering geology

Railroads
Powell, John W., 1886b

Algeria

Economic geology

Mineral resources
Gautier, E. F., 1905

Hydrogeology
Hinton, Richard J., 1890
U. S. Department of Agriculture, 1890

Alps Mountains

Geomorphology

Fluvial features, Lacustrine features
Penck, Albrecht, 1905

Glacial geology

Glacial features
Penck, Albrecht, 1905

Andes Mountains

Areal geology
Gibbon, Lardner, 1854
Macrae, Archibald, 1855

Andes Mountains (Cont'd.)
Economic geology
Gold, Mercury, Silver
Gibbon, Lardner, 1854
Geological exploration
Gibbon, Lardner, 1854
Macrae, Archibald, 1855

Antarctica
Areal geology, Geological exploration
Nordenskiöld, Otto, et al, 1904
Geomorphology
Landform description
Arctowski, Henryk, 1902
Glacial geology
Arctowski, Henryk, 1902

Appalachian Mountains
Areal geology
Emmons, S. F., 1893
McGee, W. J., et al, 1893

Arachnida
Colorado
Scudder, Samuel H., 1883

Arctic
Geomorphology
Landform description
Hall, Charles F., 1879
Glacial geology
Ice
Hall, Charles F., 1879
Paleontology
Emerson, Benjamin K., 1879

Petrology
Rock description
Emerson Benjamin K., 1879

Areal geology. See under appropriate area terms.

Argentina
Areal geology, Geological exploration
Macrae, Archibald, 1855
Economic geology
Mineral resources
Bureau of the American Republics, 1903
Geomorphology
Landform description
Bureau of the American Republics, 1903
Surveys
Land and marine surveys
Wheeler, George M., 1885

Arizona
Areal geology
Antisell, Thomas, 1855
Beale, Edward F., 1858
Beale, Edward F., &
  Engle, F. E., 1860
Blake, William P., 1901
Blake, William P., 1903
Brodie, Alexander O., 1904
Conrad, T. A., 1857
Emmons, S. F., 1893
Emory, William H., 1855
Emory, William H., 1857
Hall, James, 1857
Ives, Joseph C., 1861
Michler, N., 1857b
Newberry, J. S., 1861

Parke, John G., 1855a
Parry, C. C., &
   Schott, Arthur, 1857
Simpson, J. H., 1850b
Sitgreaves, L., 1853
Whipple, A. W., 1851

Economic geology

Coal

Bannon, M., &
   Walcott, Charles D., 1884

Mineral resources

Browne, J. R., 1867
Browne, J. R., 1868
Fergusson, D., 1863
Raymond, Rossiter W., 1869
Raymond, Rossiter W., 1870
Raymond, Rossiter W., 1871
Raymond, Rossiter W., 1872
Raymond, Rossiter W., 1873
Raymond, Rossiter W., 1874
Raymond, Rossiter W., 1875
Raymond, Rossiter W., 1877

Engineering geology

Railroads

Parke, John G., 1855a

Roads

Fergusson, D., 1863

Geological exploration

Adams, Samuel, 1870
Antisell, Thomas, 1855
Beale, Edward F., 1858
Beale, Edward F., &
   Engle, F. E., 1860
Emory, William H., 1855
Emory, William H., 1857

Fergusson, D., 1863
Gilbert, G. K., 1872
Ives, Joseph C., 1858
Ives, Joseph C., 1861
Lockwood, Daniel W., 1872
Lyle, D. A., 1872
Michler, N., 1857b
Newberry, J. S., 1861
Parke, John G., 1855a
Parry, C. C., 1869
Parry, C. C., &
   Schott, Arthur, 1857
Powell, John W., 1872a
Powell, John W., 1872b
Powell, John W., 1873
Powell, John W., 1877
Simpson, J. H., 1850b
Sitgreaves, L., 1853
Wheeler, George M., 1872a
Wheeler, George M., 1872b
Wheeler, George M., 1873
Wheeler, George M., 1874
Wheeler, George M., 1875
Wheeler, George M., 1876
Wheeler, George M., 1877
Wheeler, George M., 1878
Wheeler, George M., 1879a
Whipple, A. W., 1851

Geomorphology

Landform description

Lacey, John F., 1906
Powell, John W., 1878b

Soils, Salt River Valley

Means, Thomas H., 1901

Soils, Solomonsville

Lapham, Macy H., &
   Neill, N. P., 1904

Soils, Yuma

Holmes, J. Garnett, 1903b
Holmes, J. Garnett, et al, 1905b

Paleontology

Invertebrata

Conrad, T. A., 1857

Arizona (Cont'd.)

Hall, James, 1857
Newberry, J. S., 1861

Paleobotany

Lacey, John F., 1904
Lacey, John F., 1906
Ward, Lester F., 1901

Stratigraphy

Blake, William P., 1903
Brodie, Alexander O., 1904

Arkansas

Areal geology

Beale, Edward F., &
 Engle, F. E., 1860
Featherstonhaugh, George
 W., 1835
Loughridge, R. H., 1884b
Simpson, J. H., &
 Marcy, R. B., 1850
Weed, Walter H., 1902

Earthquakes
1811

Linn, L. F., &
 Sevier, A. H., 1836

Economic geology
Iron

Linn, L. F., &
 Sevier, A. H., 1836

Geological exploration

Beale, Edward F., &
 Engle, F. E., 1860
Featherstonhaugh, George
 W., 1835
Marcy, Randolph B., 1850
Simpson, J. H., 1850a

Geomorphology

Fluvial features, Landform description, Loess

Little, George, 1882

Landform description

Loughridge, R. H., 1884b

Log dams, Big Black River, St. Francis River, White River

Linn, L. F., &
 Sevier, A. H., 1836

Log dams, Red River

Linnard, T. B., 1844
Long, S. H., 1841
Paxton, Joseph, 1829
Shreve, Henry M., 1833
Shreve, Henry M., &
 Gratiot, C., 1834

Soils

Loughridge, R. H., 1884b

Soils, Fayetteville

Wilder, Henry J., &
 Shaw, Charles F., 1908

Soils, Miller Co.

Martin, J. O., &
 Carr, E. P., 1904b

Soils, Prairie County

Carter, William T., Jr.,
 et al, 1908

Soils, Stuttgart

Lapham, J. E., 1903

Springs, Thermal springs

Haywood, J. K., 1902
Weed, Walter H., 1902

Hydrogeology

Artesian wells

Owen, David D., 1868

Stratigraphy

Quaternary

Little, George, 1882

Artesian waters and wells. See also Springs; Hydrogeology; Groundwater; Thermal springs; Geysers.

    Algeria, Australia, California, Colorado, Great Plains, Kansas, Nebraska, New Mexico, North Dakota, South Dakota, Texas, Utah, Wyoming

        U. S. Department of Agriculture, 1890

    Algeria, Australia, Great Plains, India, Rocky Mountains

        Hinton, Richard J., 1890

    Arkansas

        Owen, David D., 1868

    Colorado, Great Plains, Kansas, Nebraska, Oklahoma

        Gregory, J. W., 1892

    Great Plains

        Hay, Robert, 1892
        Nettleton, C. E., 1892

    Nebraska

        Hicks, L. E., 1892

    New Mexico

        Hill, Robert T., 1862
        Pope, John, 1860

    Rocky Mountains

        Nettleton, C. E., 1892

    South Dakota

        Coffin, Frederick B., 1892
        Culvert, Garry E., 1892

    Texas

        Hill, Robert T., 1892
        Pope, John, 1858

Arthropoda

    Arachnida

        Colorado

            Scudder, Samuel H., 1883

    Carboniferous, Crustacea

        Packard, A. S., 1886a
        Packard, A. S., 1886b
        Packard, A. S., 1886d

    Colorado, Tertiary

        Scudder, Samuel H., 1885

    Crustacea

        Packard, A. S., 1886c

    Insecta

        Colorado

            Scudder, Samuel H., 1883

    Xiphosura

        Packard, A. S., 1886b

Associations

    Fifth International Geologic Congress

        Congrés Geologique International, 1893

    International Seismological Association

        Garland, G., 1905

Atlantic Ocean

    Marine geology

        Bottom features

            Packard, A. S., 1905

        Continental shelf

            Hilgard, J. E., 1885

        Foraminifera, Sediments

            Pourtales, L. F., 1859

        Sediments

            Agassiz, Louis, 1872b

Atlantic Ocean (Cont'd.)

Bailey, J. W., 1856a
Bailey, J. W., 1856b
Pourtales, L. F., 1854
Pourtales, L. F., 1872

Structural geology
Continental drift
Packard, A. S., 1905

Australia

Economic geology
Mining law
Raymond, Rossiter W., 1869

Geomorphology
Landform description
Thomson, J. P., 1898

Geysers
Peale, A. C., 1883

Hydrogeology
Hinton, Richard J., 1890
U. S. Department of Agriculture, 1890

Austria

Surveys
Geological & topographical surveys
Comstock, C. B., 1876
Land & marine surveys
Wheeler, George M., 1885

Aves

New Mexico, Tertiary
Cope, E. D., 1875

B

Beaches. See Shorelines; Coastal features under Geomorphology; Shore features under Marine geology.

Belgium

Surveys
Geological & topographical surveys
Comstock, C. B., 1876
Land & marine surveys
Wheeler, George M., 1885

Bibliography

Earthquakes
Iceland
Boehmer, George H., 1886

Geodesy
Gore, James H., 1903

Invertebrate paleontology
North America, 1884
Marcou, John B., 1885
North America, 1885
Marcou, John B., 1886
North America, 1886
Marcou, John B., 1889

Mineralogy
1885
Dana, Edward S., 1886
1886
Dana, Edward S., 1889

Paleontology
1887
Williams, Henry S., 1890
1888
Williams, Henry S., 1890

Petrology
1887
Merrill, George P., 1890
1888
Merrill, George P., 1890
Seismology
1884
Rockwood, Charles G., Jr., 1885
1885
Rockwood, Charles G., Jr., 1886
1886
Rockwood, Charles G., Jr., 1889
Volcanology
1884
Rockwood, Charles G., Jr., 1885
1885
Rockwood, Charles G., Jr., 1886
1886
Rockwood, Charles G., Jr., 1886
Iceland
Boehmer, George H., 1886

Biographical sketch. See **Memoirs**

Biography. See **Memoirs**; see **Memoirs** under **History**; see **Memoirs** under appropriate discipline.

Biology
   Relation to geology
      Lapworth, Charles, 1904
      White, Charles A., 1893

Bolivia
   Areal geology, Geological exploration
      Gibbon, Lardner, 1854
   Economic geology
      Gold, Mercury, Silver
      Gibbon, Lardner, 1854
      Mineral resources
      Bureau of the American Republics, 1904a
   Geomorphology
      Landform description
      Bureau of the American Republics, 1904a
   Surveys
      Land and marine surveys
      Wheeler, George M., 1885

Brazil
   Areal geology, Geological exploration
      Herndon, Lewis, 1853
   Economic geology
      Copper, Diamonds, Gold Salt, Silver
      Herndon, Lewis, 1853
      Diamonds
      Derby, Orville A., 1907
      Gems
      Bureau of the American Republics, 1905b
      Mineral resources
      Bureau of the American Republics, 1902
   Geomorphology
      Landform description
      Bureau of the American Republics, 1902

British Columbia
Areal geology, Geological exploration
Blake, Theodore A., 1869
Economic geology
Railroads
Powell, John W., 1886b
Geomorphology
Landform description
Blake, Theodore A., 1869

C

Calderas. See Volcanoes.

California
Areal geology
Abbot, Henry L., 1855
Antisell, Thomas, 1855
Ashburner, William, 1867
Blake, William P., 1855a
Blake, William P., 1855b
Blake, William P., 1856b
Bowers, Stephen, 1878
Bowman, Amos, 1875
Brown, Edwin C., 1905
Chandler, M. T. W., 1857
Conrad, T. A., 1857
Derby, George H., 1850
Derby, George H., 1852b
Emory, William H., 1848
Emory, William H., 1857
Fremont, John C., 1845
Fremont, John C., 1848
Hall, James, 1857
Hilgard, Eugene W., 1884b
Ives, Joseph C., 1858
Ives, Joseph C., 1861
Johnston, A. R., 1858
Manson, Marsden, 1902
Marcou, Jules, 1855a
Marcou, Jules, 1855c
Marcou, Jules, 1876
Michler, N., 1857b
Newberry, J. S., 1855
Newberry, J. S., 1861
Parke, John G., 1855a
Parry, C. C., &
 Schott, Arthur, 1857
Preston, E. B., 1905
Raymond, Rossiter W., 1875
Sitgreaves, L., 1853
Tyson, P. T., 1850
Whipple, A. W., 1851
Whipple, A. W., 1855
Williamson, R. S., 1855

Earthquakes
1854
Bache, A. D., 1856a
1868
Hilgard, J. E., 1872

Economic geology
Coal
Frazer, John F., 1850
Gabb, W. M., 1867
Gold
Ashburner, William, 1867
Bowman, Amos, 1873
King, T. Butler, 1850
Mendell, G. H., 1882
Waldeyer, Charles, 1873
Limestone
Williamson, R. S., 1850
Mineral resources
Brown, Edwin C., 1905
Browne, J. R., 1867
Browne, J. R., 1868
Preston, E. B., 1905
Raymond, Rossiter W., 1870
Raymond, Rossiter W., 1871
Raymond, Rossiter W., 1872

Raymond, Rossiter W., 1873
Raymond, Rossiter W., 1874
Raymond, Rossiter W., 1875
Raymond, Rossiter W., 1877
Tyson, P. T., 1850

Sierra Nevada

Bowman, Amos, 1875

Engineering geology

Railroads

Abbot, Henry L., 1855
Beckwith, E. G., 1855a
Parke, John G., 1855a
Schiel, James, 1855a
Whipple, A. W., 1855
Williamson, R. S., 1855

Rivers, Waste disposal

Mendell, G. H., 1882

Geological exploration

Abbot, Henry L., 1855
Adams, Samuel, 1870
Antisell, Thomas, 1855
Beckwith, E. G., 1855a
Blake, William P., 1855a
Blake, William P., 1855b
Blake, William P., 1856b
Chandler, M. T. W., 1857
Conkling, A. R., 1877a
Conkling, A. R., 1878
Derby, George H., 1850
Derby, George H., 1852b
Emory, William H., 1848
Emory, William H., 1857
Fremont, John C., 1845
Fremont, John C., 1848
Gilbert, G. K., 1872
Ives, Joseph C., 1858
Ives, Joseph C., 1861
Johnston, A. R., 1858
King, Clarence, 1873
King, Clarence, 1874

King, Clarence, 1876
King, Clarence, 1877
Lyle, D. A., 1872
Marcou, Jules, 1855a
Marcou, Jules, 1855c
Marcou, Jules, 1876
Michler, N., 1857b
Newberry, J. S., 1855
Newberry, J. S., 1861
Parke, John G., 1855a
Parry, C. C., &
  Schott, Arthur, 1857
Schiel, James, 1855a
Sitgreaves, L., 1853
Wheeler, George M., 1874
Wheeler, George M., 1875
Wheeler, George M., 1876
Wheeler, George M., 1877
Wheeler, George M., 1878
Wheeler, George M., 1879a
Whipple, A. W., 1851
Whipple, A. W., 1855
Williamson, R. S., 1855

Geomorphology

Alkaline lakes

Loew, Oscar, 1876a

Drainage, Erosion, Fluvial features

Mendell, G. H., 1882

Drainage, San Joaquin River

Soulé, Frank, 1902

Drainage, Yuba River

Manson, Marsden, 1902

Fluvial features, Tertiary

Bowman, Amos, 1873

Geysers

Peale, A. C., 1883

Landform description

Hilgard, Eugene W., 1884b
Loew, Oscar, 1876b
Mendell, G. H., 1882
Powell, John W., 1878b
Soulé, Frank, 1902

California (Cont'd.)

Soils
Hilgard, Eugene W., 1884b

Soils, Bakersfield
Lapham, Macy H., &
 Jensen, Charles A., 1905

Soils, Fresno
Means, Thomas H., &
 Holmes, J. Garnett, 1901

Soils, Hanford
Lapham, Macy H., &
 Heileman, W. H., 1902a

Soils, Imperial
Holmes, J. Garnett,
 et al, 1904a
Means, Thomas H., &
 Holmes, J. Garnett, 1902

Soils, Indio
Holmes, J. Garnett, et al,
 1904b

Soils, Los Angeles
Mesmer, Louis, 1904

Soils, Sacramento
Lapham, Macy H., &
 Mackie, W. W., 1905

Soils, Salinas Valley
Lapham, Macy H., &
 Heileman, W. H., 1902b

Soils, San Bernadino Valley
Holmes, J. Garnett,
 et al, 1905a

Soils, San Gabriel
Holmes, J. Garnett, 1902

Soils, San Jose
Lapham, Macy H., 1904

Soils, Santa Ana
Holmes, J. Garnett, 1901

Soils, Stockton
Lapham, Macy H., &
 Mackie, W. W., 1907

Soils, Ventura
Holmes, J. Garnett, &
 Mesmer, Louis, 1902

Soils, Yuma
Holmes, J. Garnett,
 et al, 1905b

Thermal springs
Loew, Oscar, 1876a

Hydrogeology
Hinton, Richard J., 1892
U. S. Department of Agriculture, 1890

Salinas Valley
Nutter, Edward H., 1902

Marine geology

Tsunamis
Bache, A. D., 1856a
Hilgard, J. E., 1872

Mineralogy
Blake, William P., 1867

Mineral waters
Loew, Oscar, 1876a

Paleontology

Diatoms
Bailey, J. W., 1855b

Invertebrata
Conrad, T. A., 1855a
Conrad, T. A., 1855b
Conrad, T. A., 1855d
Conrad, T. A., 1857
Hall, James, 1845a
Hall, James, 1855
Hall, James, 1857
Newberry, J. S., 1861
Schiel, James, 1855b

Man, fossil
Holmes, William H., 1901

Paleobotany
Bailey, J. W., 1855a
Schaeffer, George C., 1855
Vertebrata, Pisces
Agassiz, Louis, 1855
Stratigraphy
Hall, James, 1845b

Cambrian
Evolution of life
Brooks, William K., 1896
Rocky Mountains
Stratigraphy
Hayden, F. V., 1878c

Canada. See also under Province names.
Economic geology
Copper, Lake Superior region
Locke, John, 1848
Mining law
Raymond, Rossiter W., 1869

Canals. See under Engineering geology.

Carboniferous
Crustacea
Packard, A. S., 1886a
Packard, A. S., 1886b
Packard, A. S., 1886d
Great Plains
Paleontology
Shumard, B. F., 1853a
Oklahoma
Paleontology
Shumard, B. F., 1853a

Texas
Paleontology
Shumard, B. F., 1853a
United States
Hitchock, C. H., 1874

Caribbean region. See also Caribbean Sea.
Economic geology
Mineral resources
U. S. War Department, 1906a
U. S. War Department, 1906b
Geomorphology
Landform description
U. S. War Department, 1906a
U. S. War Department, 1906b
Landform description, Isle of Pines
Hayes, Willard, 1906
Volcanology
Hill, Robert T., 1905b
Guadeloupe, Martinique, Saba
Hovey, Edmund O., 1905a
Martinique
Heilprin, Angelo, 1905a
Heilprin, Angelo, 1905b
Martinique, St. Vincent
Anderson, Tempest, & Flett, John S., 1903
Russell, Israel C., 1903
Pelee
Bauer, L. A., 1905
St. Kitts, St. Vincent, Statia
Hovey, Edmund O., 1905b

Caribbean Sea
  Marine geology
    Bottom features, Sediments
      Agassiz, Alexander, 1881

Caves. See also Solution features under Geomorphology.
  Features
    South Dakota
      Gamble, Robert J., 1902
    United States
      Martel, E. A., 1905

Cement materials. See Limestone.

Central America. See also names of individual countries.
  Areal geology
    Davis, C. H., 1866
  Engineering geology
    Canals, Railroads
      Davis, C. H., 1866
  Surveys
    Land and marine surveys
      Wheeler, George M., 1885

Changes in level. See also Uplifts; Isostatic rebound under Glaciation.
  Causes
    Davis, W. M., 1905a
    LeConte, Joseph, 1898
  Causes, Cycles, Rates
    Blytt, A., 1890
  Continents
    Gilbert, G. K., 1893

Great Lakes region
  Isostatic rebound
    Gilbert, G. K., 1899
Maine, Massachusetts, Nova Scotia
  Isostatic rebound
    Mitchell, Henry, 1880
Ontario, Quebec
  Isostatic rebound
    Bell, Robert, 1898
Shore features
  Massachusetts
    Gulliver, F. P., 1905

Chemical analysis
  Soils
    Cameron, Frank K., 1900

Chile
  Areal geology, Geological exploration
    Gilliss, J. M., 1855
    Macrae, Archibald, 1855
  Earthquakes
    1570 to 1851
      Gilliss, J. M., 1855
  Geomorphology
    Geysers
      Peale, A. C., 1883
    Mineral waters
      Gilliss, J. M., 1855
      Smith, J. Lawrence, 1855
  Mineralogy
    Meteorites
      Philippi, R. A., 1855
  Paleontology
    Invertebrata
      Conrad, T. A., 1855c

Index

Vertebrata, Mammalia
Wyman, Jeffries, 1855
Surveys
Land and marine surveys
Wheeler, George M., 1885

China
Geomorphology
Geysers, Tibet
Peale, A. C., 1883
Landform description, Tibet
Younghusband, Frank, 1906

Climates, ancient. See **Paleoclimatology.**

Coal
Alabama
Alabama, University of, 1902
Alaska
Barnard, E. C., 1899
Brooks, Alfred H., 1899
Eldridge, G. H., 1898a
Eldridge, G. H., &
 Muldrow, Robert, 1899
Mendenhall, Walter C., 1899
Peters, W. J., &
 Brooks, Alfred H., 1899
Schrader, Frank C., 1899
Arizona
Bannon, M., &
 Walcott, Charles D., 1884
California
Frazer, John F., 1850
Colorado
Lesquereux, L., 1873

Great Plains
Hodge, Joseph T., 1872
Marcy, Randolph B., 1853
Illinois
Locke, John, 1844
Owen, David D., 1840
Illinois, Indiana, Iowa, Kentucky
Owen, David D., 1844
Illinois, Kentucky, Missouri, Ohio, Pennsylvania, West Virginia
Jamison, J., 1853
Kentucky
Shaler, N. S., 1875
Kentucky, Virginia, West Virginia
Brown, C. Newton, 1900
Montana, Ohio, Washington
Pumpelly, Raphael, 1886
Nevada, New Mexico, Utah, Wyoming
Lesquereux, L., 1873
New Mexico
Sheridan, Jo E., 1903
Sheridan, Jo E., 1904
North Carolina
Laidley, T. T. S., 1856
Wilkes, Charles; Hunt, H., &
 Martin, D. B., 1859
Nova Scotia
Hamilton, Pierce S., 1868
Oklahoma, Texas
Marcy, Randolph B., 1853
Origin
LeConte, Joseph, 1858
U. S. Department of Commerce and Labor, 1905

Coal (Cont'd.)

Panama
Engle, F., 1860
Evans, John, 1860

Philippines
Smith, Warren D., 1905a
Smith, Warren D., 1905b
U. S. War Department, 1905
Wigmore, H. L., 1904
Wigmore, H. L., 1905a
Wigmore, H. L., 1905b

Rocky Mountains
Hodge, Joseph T., 1872
Lesquereux, L., 1873

South Dakota
Taylor, James W., 1867
Taylor, James W., 1868

Taiwan
Jones, George, 1856c

United States
Hitchock, C. H., 1874
Pumpelly, Raphael, 1886

United States, Western
Hayden, F. V., 1868

Collections

Building stones
U. S. National Museum
Merrill, George P., 1889

Gems
U. S. National Museum
Kunz, George F., 1889
Tassin, Wirt, 1902a

Meteorites
U. S. National Museum
Clarke, F. W., 1889
Tassin, Wirt, 1902b

Yale University
Brush, G. J., 1869

Minerals
Land Office
Roessler, A. R.; Wilson, Joseph S., & Goodwin, J. R., 1871
Library of Congress
Hughes, George W., et al, 1837
U. S. National Museum
Merrill, George P., 1891
Tassin, Wirt, 1897
Tassin, Wirt, 1899a
Tassin, Wirt, 1899b

Non-metallic minerals
U. S. National Museum
Merrill, George P., 1901

Ornamental stones
U. S. National Museum
Merrill, George P., 1889

Rocks
U. S. National Museum
Merrill, George P., 1891

Colombia

Areal geology
Davis, C. H., 1866
Selfridge, Thomas O., 1871

Economic geology
Salt
Bureau of the American Republics, 1905a

Engineering geology
Canals
Collins, Frederick, 1879
Davis, C. H., 1866
Michler, N., 1861

Selfridge, Thomas O., 1871
Selfridge, Thomas O., 1874
Railroads
Davis, C. H., 1866
Geological exploration
Maack, G. A., 1874
Michler, N., 1861
Schott, Arthur, 1861b
Selfridge, Thomas O., 1874
Geomorphology
Landform description
Collins, Frederick, 1879
Schott, Arthur, 1861b
Selfridge, Thomas O., 1874
Structural geology
Schott, Arthur, 1861a

Colorado
Areal geology
Emmons, S. F., 1893
Gregory, J. W., 1892
Hayden, F. V., 1883a
Loew, Oscar, 1875
McCauley, C. A. H., 1879
Economic geology
Coal
Lesquereux, L., 1873
Gold, Silver
Taylor, James W., 1867
Taylor, James W., 1868
Mineral resources
Elliott, R. S., 1872
Hawn, F., 1874d
McCauley, C. A. H., 1878
Raymond, Rossiter W., 1869
Raymond, Rossiter W., 1870
Raymond, Rossiter W., 1871

Raymond, Rossiter W., 1872
Raymond, Rossiter W., 1873
Raymond, Rossiter W., 1874
Raymond, Rossiter W., 1875
Raymond, Rossiter W., 1877
Geological exploration
Adams, Samuel, 1870
Campbell, Donald, 1874
Conkling, A. R., 1877b
Hawn, F., 1874a
Hawn, F., 1874b
Hawn, F., 1874c
Hawn, Laurens, 1874
Hayden, F. V., 1872b
Hayden, F. V., 1877
Hayden, F. V., 1878b
Hayden, F. V., 1883a
King, Clarence, 1871
King, Clarence, 1873
King, Clarence, 1874
King, Clarence, 1876
King, Clarence, 1877
McCauley, C. A. H., 1878
Parry, C. C., 1869
Peale, A. C., 1873
Powell, John W., 1872a
Ruffner, E. H., 1874
Stevenson, J. T., 1879
Warren, G. K., 1858
Wheeler, George M., 1873
Wheeler, George M., 1874
Wheeler, George M., 1875
Wheeler, George M., 1876
Wheeler, George M., 1877
Wheeler, George M., 1878
Wheeler, George M., 1879a
Geomorphology
Landform description
McCauley, C. A. H., 1878
McCauley, C. A. H., 1879
Powell, John W., 1878b

Colorado (Cont'd.)

Soils, Arkansas Valley
Lapham, Macy H.,
  et al, 1903

Soils, Cache a la Poudre Valley
Means, Thomas H., 1900a

Soils, Grand Junction
Holmes, J. Garnett, &
  Rice, Thomas D., 1907

Soils, Greeley
Holmes, J. Garnett, &
  Neill, N. P., 1905

Soils, San Luis Valley
Holmes, J. Garnett, 1904

Thermal springs
McCauley, C. A. H., 1879
Smart, Charles, 1879

Hydrogeology
Gregory, J. W., 1892
U. S. Department of Agriculture, 1890

Mineralogy
Loew, Oscar, 1875

Mineral waters
Smart, Charles, 1879

Paleontology
Arachnida, Insecta
Scudder, Samuel H., 1883

Arthropoda, Tertiary
Scudder, Samuel H., 1885

Invertebrata
Hayden, F. V., 1883a
Meek, F. B., 1872c
Scudder, Samuel H., 1883
Scudder, Samuel H., 1885
White, Charles A., 1883

Vertebrata, Tertiary
Cope, E. D., 1884

Petrology
Conkling, A. R., 1877c

Colorado Plateau, see also names of appropriate states

Areal geology
Ives, Joseph C., 1861
Newberry, J. S., 1861

Geological exploration
Geikie, Archibald, 1884
Ives, Joseph C., 1861
Newberry, J. S., 1861
Powell, John W., 1872a
Powell, John W., 1872b
Powell, John W., 1873

Paleontology
Invertebrata
Newberry, J. S., 1861

Congresses. See **Associations.**

Connecticut

Economic geology
Iron
Pumpelly, Raphael, 1886

Geomorphology
Fluvial features
Powell, Charles F., 1905

Soils, Connecticut Valley
Dorsey, Clarence W., &
  Bonsteel, Jay A., 1900
Fippin, Elmer O., 1904

Marine geology
Continental shelf
Lindenkohl, A., 1885

Index

Construction materials
  Chemistry, Description, Microscopic structure
    Hawes, George W., 1884
  Collections
    U. S. National Museum
    Merrill, George P., 1889
  History, Occurrences, Production, Properties
    Merrill, George P., 1889
  Nicaragua
    Powell, John W., 1886a
  Philippines
    Ickis, E. M., &
      Field, E. M., 1905
  Quarries and quarry regions
    Hawes, George W., 1884
    Merrill, George P., 1889

Continental drift
  Animal distribution
    Africa, Brazil
      Packard, A. S., 1905
  Bottom features
    Atlantic Ocean
      Packard, A. S., 1905
  Crustal structure
    Lineation
      Hobbs, William H., 1905

Continental shelf
  Atlantic Ocean
    United States
      Hilgard, J. E., 1885
      Lindenkohl, A., 1885

Continents. See also **Continental drift.**
  Changes in level, Growth
    Gilbert, G. K., 1893
  Distribution, Permanence
    Gregory, J. W., 1899
  Origin
    Kelvin, William T., 1898
    Murray, John, 1901

Copper
  Alaska
    Peters, W. J., &
      Brooks, Alfred H., 1899
    Schrader, Frank C., 1899
    Schrader, Frank C., &
      Spencer, Arthur C., 1901
  Brazil
    Herndon, Lewis, 1853
  England
    Bell, William H., 1844
  England, Cornwall, Devon
    Hughes, George W., 1844
  Great Plains
    Marcy, Randolph B., 1853
  Iowa
    Owen, David D., 1848a
  Lake Superior region
    Allen, J., 1834
    Bell, William H., 1844
    Burt, William A., 1846
    Cass, Lewis, 1825
    Channing, William F., 1848a
    Channing, William F., 1848b
    Channing, William F., 1848c

Copper (Cont'd.)

Foster, J. W., 1848
Foster, J. W., &
 Whitney, J. D., 1850b
Gray, A. B., 1846
Houghton, Douglas, 1832
Hubbard, Bela, 1846
Jackson, Charles T., 1848a
Jackson, Charles T., 1848b
Jackson, Charles T., 1849
Owen, David D., 1848a
Stockton, John, et al, 1845
Whitney, J. D., 1848a
Whitney, J. D., 1848b

Michigan

Channing, William F., 1848a
Channing, William F., 1848b
Channing, William F., 1848c
Foster, J. W., &
 Whitney, J. D., 1850b
Jackson, Charles T., 1848a
Whitney, J. D., 1848a

Montana

Taylor, James W., 1867
Taylor, James W., 1868

New Mexico

Taylor, James W., 1867
Taylor, James W., 1868

Oklahoma, Texas

Marcy, Randolph B., 1853

Wisconsin

Bell, William H., 1844
Locke, John, 1844
Owen, David D., 1840
Owen, David D., 1848a

Core

Structure

Hennessey, Henry, 1891
Peirce, Benjamin, 1881

Correlation

Methods

Fossils

White, Charles A., 1893

Paleozoic

Hall, James, 1851c

Costa Rica

Earthquakes

Jones, James O., 1903

Structural geology

Orogeny, Uplifts

Sapper, Karl, 1905

Volcanology

Jones, James O., 1903
Sapper, Karl, 1905

Cretaceous

Great Plains, Oklahoma, Texas

Paleontology

Shumard, B. F., 1853a

New Mexico

Paleontology

Abert, J. W., 1848
Bailey, J. W., 1848

Crust

Density, Movements, Structure

Gilbert, G. K., 1893

Movements

Blytt, A., 1890
LeConte, Joseph, 1898

Origin, Temperature

Pilar, George, 1877

Structure
Gregory, J. W., 1899
Peirce, Benjamin, 1881
Temperature
Schott, Charles C., 1875

Crustacea
Carboniferous
Packard, A. S., 1886a
Carboniferous, Malacostraca
Packard, A. S., 1886d
Carboniferous, Xiphosura
Packard, A. S., 1886b
Gampsonychidae, Paleocaris typus
Packard, A. S., 1886c

Cryolite. See **Fluorspar**.

Crystal structure. See also **Crystallography**.
Brezina, Aristides, 1873
Gold
Blake, William P., 1885
Packing
Liveing, G. D., 1893

Crystallography. See also **Crystal structure**.
Brezina, Aristides, 1873
Crystal growth
Judd, John W., 1893
Crystal growth, Internal structure
Liveing, G. D., 1893

D

Deformation. See also **Faults; Structural geology**.
Experimental studies
Hallock, William, 1893

Delaware
Geomorphology
Deltas, Fluvial features
Mitchell, Henry, 1887b
Fluvial features
Marinden, Henry L., 1883
Marinden, Henry L., 1885
Mitchell, Henry, 1889a
Shore features, Fluvial features
Mitchell, Henry, 1884
Soils, Dover
Bonsteel, F. E., &
Ayrs, O. L., 1904
Marine geology
Continental shelf
Lindenkohl, A., 1885
Shore features
Mitchell, Henry, 1884
Mitchell, Henry, 1889a

Deltas. See also under **Geomorphology**.
Evolution
Mississippi delta
U. S. Army Corps of Engineers, 1874
Geomorphology
Delaware River
Mitchell, Henry, 1887b

Deltas (Cont'd.)

Mississippi delta
Delafield, Richard, 1829
Ellet, Charles, Jr., 1851
Graham, George, 1829
Long, S. H., &
 Humphreys, A. A., 1851

Sedimentation
Mississippi delta
Ellet, Charles, Jr., 1851
Ellet, Charles, Jr., 1852
Long, S. H., &
 Humphreys, A. A., 1851

Sediments
Mississippi delta
Forshey, C. G., 1875

Special features
Mudlumps
Delafield, Richard, 1829
Ellet, Charles, Jr., 1851

Denmark

Surveys
Land and marine surveys
Wheeler, George M., 1885

Deposition. See **Sedimentation**.

Diamonds. See also **Gems**.

Babinet, Jacques, 1871

Brazil
Derby, Orville A., 1907
Herndon, Lewis, 1853

Description
Cullinan diamond
Hatch, F. H., &
 Corstorphine, G. S., 1906

Origin
Crookes, William, 1898
Williams, Gardner F., 1906

Origin, Properties
Crookes, William, 1898

Paraguay
Gibbon, Lardner, 1854

United States
Origin
Hobbs, William H., 1902

Diatoms

California
Bailey, J. W., 1855b

Louisiana, Recent
Hilgard, Eugene W., &
 Hopkins, F. V., 1878

Oregon
Bailey, J. W., 1845

Dinosauria. See **Reptilia**.

Dominican Republic

Areal geology
Wade, B. F.; White, A. D.,
 & Howe, S. G., 1871

Economic geology
Mineral resources
Blake, William P., 1871c
Marvine, A. R., 1871
McClellan, George B., 1871

Petroleum
Siguel, F., 1871

Geological exploration
Adam, J. S., 1871
Blake, William P., 1871a
Blake, William P., 1871b

Geomorphology
Landform description
McClellan, George B., 1871

E

Earth. See also **Geodesy; Core; Crust; Magnetic field.**

Age
Chamberlain, T. C., 1901
Geikie, Archibald, 1893
Joly, J., 1901
Kelvin, William T., 1898
King, Clarence, 1894
Walcott, Charles D., 1894

Chemistry
Hunt, T. Sterry, 1871

Core, Structure. See also under **Core.**
Hennessey, Henry, 1891

Crust, Shape. See also under **Crust.**
Gregory, J. W., 1899

Elasticity, Shape, Structure
Chree, C., 1893

Electricity
Matteucci, Carlo, 1871

Evolution
Prestwich, Joseph, 1876
Sollas, W. J., 1901

History
Sea level changes
Blytt, A., 1890

History, Origin, Temperature
Pilar, George, 1877

Interior
Peirce, Benjamin, 1881

Magnetism
Creak, Ettrick W., 1904

Origin
Woodward, Robert S., 1891

Physical properties
Density
Wilsing, J., 1890

Rotation
Stream deflection
Gilbert, G. K., 1885

Shape
Merino, Miguel, 1864

Temperature
King, Clarence, 1894
Schott, Charles C., 1875

Earthquakes. See aslo **Seismology**; see also **Earthquakes** under appropriate area terms.
Henry, Joseph, 1856
Mallet, R., 1860

Arkansas
1811
Linn, L. F., &
Sevier, A. H., 1836

California
1854
Bache, A. D., 1856a
1868
Hilgard, J. E., 1872

Chile
1570 to 1851
Gilliss, J. M., 1855

Costa Rica, Guatemala, Nicaragua
Jones, James O., 1903

Distribution
Lalleman, G., 1905

Guatemala
1857, 1858
Canudas, A., 1859

Earthquakes (Cont'd.)

Iceland
Boehmer, George H., 1886
Stefansson, Jon, 1907

Japan
1854
Bache, A. D., 1856a

Mexico
1866
Sartorius, Charles, 1867

Missouri
1811
Dudley, Timothy, 1859

Nicaragua
Dutton, C. E., 1902
Wheeler, E. S., 1902

North Carolina
1874
du Pre, Warren, 1875

Panama
1882
Nicaragua Canal Commission, 1906

Peru
1868
Campbell, John V., 1871

Turkey
1881
Rittenhouse, H. O.; Knight, A. M., & Huse, H. P., 1881

Economic geology. See also Mineral deposits, genesis.

Education

Berlin
Ward, Willard P., 1869

Clausthal, Freiberg, Paris, Prussia
Raymond, Rossiter W., 1869

Freiberg
Lyman, Benjamin S., 1869

Methods

Cement manufacture
Eckel, Edwin C., 1903

Nevada
Church, John A., 1877
Church, John A., 1878
Wright, H. G.; Foster, J. G., & Newcomb, Wesley, 1872

Resources

Alabama
Walcott, Charles D., 1901
Walcott, Charles D., 1905

Alaska
Abercrombie, W. R., 1900
Blake, Theodore A., 1869
Browne, J. R., 1867
Browne, J. R., 1868
Bryant, Charles, 1870
Bulkley, Charles S., 1868
Davidson, George, 1868
Eldridge, G. H., 1898b
Eldridge, G. H., 1899
McIntyre, H. H., 1870
Mendenhall, Walter C., 1898a
Mendenhall, Walter C., 1898b
Mendenhall, Walter C., 1898c
Morris, William G., 1879
Raymond, Charles C., 1871
Raymond, Rossiter W., 1874
Schrader, Frank C., 1898
Senate Committee on Military Affairs, 1900

Spurr, J. E., & Post, W. S., 1899
Williams, Harry G., 1870

Algeria
Gautier, E. F., 1905

Andes Mountains
Gibbon, Lardner, 1854

Argentina
Bureau of the American Republics, 1903

Arizona
Browne, J. R., 1867
Browne, J. R., 1868
Raymond, Rossiter W., 1869
Raymond, Rossiter W., 1870
Raymond, Rossiter W., 1871
Raymond, Rossiter W., 1872
Raymond, Rossiter W., 1873
Raymond, Rossiter W., 1874
Raymond, Rossiter W., 1875
Raymond, Rossiter W., 1877

Bolivia
Bureau of the American Republics, 1904a

Brazil
Bureau of the American Republics, 1902

British Columbia
Blake, Theodore A., 1869

California
Brown, Edwin C., 1905
Browne, J. R., 1867
Browne, J. R., 1868
Preston, E. B., 1905

Raymond, Rossiter W., 1869
Raymond, Rossiter W., 1870
Raymond, Rossiter W., 1871
Raymond, Rossiter W., 1872
Raymond, Rossiter W., 1873
Raymond, Rossiter W., 1874
Raymond, Rossiter W., 1875
Raymond, Rossiter W., 1877
Tyson, P. T., 1850

Canada, Nova Scotia, Saskatchewan
Taylor, James W., 1867
Taylor, James W., 1868

Caribbean region
U. S. War Department, 1906a
U. S. War Department, 1906b

Coal
LeConte, Joseph, 1858

Coal, Alabama
Alabama, University of, 1902

Coal, Alaska
Barnard, E. C., 1899
Brooks, Alfred H., 1899
Eldridge, G. H., 1898a
Eldridge, G. H., & Muldrow, Robert, 1899
Mendenhall, Walter C., 1899
Peters, W. J., & Brooks, Alfred H., 1899
Schrader, Frank C., 1899

Economic geology (Cont'd.)

Coal, Arizona
Bannon, M., & Walcott,
  Charles D., 1884

Coal, California
Frazer, John F., 1850
Gabb, W. M., 1867

Coal, Colorado
Lesquereux, L., 1873

Coal, Great Plains
Hodge, Joseph T., 1872
Marcy, Randolph B., 1853

Coal, Illinois
Jamison, J., 1853
Locke, John, 1844
Owen, David D., 1840

Coal, Indiana, Illinois, Iowa
Owen, David D., 1844

Coal, Kentucky
Brown, C. Newton, 1900
Owen, David D., 1844
Shaler, N. S., 1875
Jamison, J., 1853

Coal, Missouri, Kentucky

Coal, Nevada, New Mexico
Lesquereux, L., 1873

Coal, New Mexico
Sheridan, Jo E., 1903
Sheridan, Jo E., 1904

Coal, North Carolina
Laidley, T. T. S., 1856
Wilkes, Charles; Hunt, H., &
  Martin, D. B., 1859

Coal, Nova Scotia
Hamilton, Pierce S., 1868

Coal, Ohio
Jamison, J., 1853

Coal, Oklahoma, Texas
Marcy, Randolph B., 1853

Coal, Oregon
Gabb, W. M., 1867

Coal, Panama
Engle, F., 1860
Evans, John, 1860

Coal, Pennsylvania
Jamison, J., 1853

Coal, Philippines
Smith, Warren D., 1905a
Smith, Warren D., 1905b
U. S. War Department, 1905
Wigmore, H. L., 1904
Wigmore, H. L., 1905a
Wigmore, H. L., 1905b

Coal, Rocky Mountains
Hodge, Joseph T., 1872
Lesquereux, L., 1873

Coal, Taiwan
Jones, George, 1856c

Coal, United States
Hitchock, C. H., 1874
Pumpelly, Raphael, 1886

Coal, Utah
Lesquereux, L., 1873

Coal, Virginia, West Virginia
Brown, C. Newton, 1900

Coal, West Virginia
Jamison, J., 1853

Coal, Western United States
Hayden, F. V., 1868

Coal, Wyoming
Lesquereux, L., 1873

Colorado
Elliott, R. S., 1872
Hawn, F., 1874d

McCauley, C. A. H., 1878
Raymond, Rossiter W., 1870
Raymond, Rossiter W., 1871
Raymond, Rossiter W., 1872
Raymond, Rossiter W., 1873
Raymond, Rossiter W., 1874
Raymond, Rossiter W., 1875
Raymond, Rossiter W., 1877
Taylor, James W., 1867
Taylor, James W., 1868

Construction materials, Nicaragua
Powell, John W., 1886a

Construction materials, Philippines
Ickis, E. M., & Field, E. M., 1905

Construction materials, United States
Hawes, George W., 1884

Copper, Alaska
Peters, W. J., & Brooks, Alfred H., 1899
Schrader, Frank C., 1899
Schrader, Frank C., & Spencer, Arthur C., 1901

Copper, Brazil
Herndon, Lewis, 1853

Copper, England
Bell, William H., 1844
Hughes, George W., 1844

Copper, Great Plains
Marcy, Randolph B., 1853

Copper, Iowa
Owen, David D., 1848a

Copper, Lake Superior region
Allen, J., 1834
Bell, William H., 1844
Burt, William A., 1846
Cass, Lewis, 1825
Channing, William F., 1848a
Channing, William F., 1848b
Channing, William F., 1848c
Foster, J. W., 1848
Foster, J. W., & Whitney, J. D., 1850b
Gray, A. B., 1846
Houghton, Douglas, 1832
Hubbard, Bela, 1846
Jackson, Charles T., 1848a
Jackson, Charles T., 1848b
Jackson, Charles T., 1849
Locke, John, 1848
Owen, David D., 1848a
Schoolcraft, Henry R., 1822
Stockton, John, et al, 1845
Whitney, J. D., 1848a
Whitney, J. D., 1848b

Copper, Michigan
Channing, William F., 1848a
Channing, William F., 1848b
Channing, William F., 1848c
Jackson, Charles T., 1848a
Whitney, J. D., 1848a

Copper, Oklahoma, Texas
Marcy, Randolph B., 1853

Copper, Wisconsin
Bell, William H., 1844
Locke, John, 1844
Owen, David D., 1840
Owen, David D., 1848a

Economic geology (Cont'd.)

Diamonds, Brazil
Derby, Orville A., 1907
Herndon, Lewis, 1853

Diamonds, Paraguay
Gibbon, Lardner, 1854

Diamonds, United States
Hobbs, William H., 1902

Dominican Republic
Blake, William P., 1871c
Marvine, A. R., 1871
McClellan, George B., 1871

Fluorspar, Greenland
Quale, Paul, 1867

Gas, natural, Indiana, Ohio
Orton, Edward, 1893

Gems, Brazil
Bureau of the American Republics, 1905b

Georgia
Howard, C. W., 1867
Taylor, James W., 1867

Gold
Blake, William P., 1869

Gold, Alaska
Barnard, E. C., 1899
Brooks, Alfred H., 1899
Brooks, Alfred H.; Richardson, George B., & Collier, Arthur J., 1901
Eldridge, G. H., & Muldrow, Robert, 1899
Emmons, S. F., 1898
Mendenhall, Walter C., 1899
Mendenhall, Walter C., 1901
Peters, W. J., & Brooks, Alfred H., 1899
Schrader, Frank C., 1899
Schrader, Frank C., & Brooks, Alfred H., 1900
Schrader, Frank C., & Spencer, Arthur C., 1901
Walcott, Charles D., 1897

Gold, Andes Mountains, Bolivia
Gibbon, Lardner, 1854

Gold, Brazil
Herndon, Lewis, 1853

Gold, California
Ashburner, William, 1867
Bowman, Amos, 1873
King, T. Butler, 1850
Mendell, G. H., 1882
Waldeyer, Charles, 1873

Gold, Nevada
Blatchly, A., 1867
Edmunds, J. M., et al, 1863

Gold, Nova Scotia
Hamilton, Pierce S., 1868

Gold, Oklahoma
Bain, H. Foster, 1904

Gold, Panama
Engle, F., 1860

Gold, Peru
Gibbon, Lardner, 1854

Gold, United States
King, Clarence; Emmons, S. F., & Becker, G. F., 1885

Great Plains
Dodge, G. M., 1868
Dodge, G. M., 1869
Mullan, John, 1863
Van Lennep, D., 1868

Gypsum, Great Plains, Oklahoma, Texas
Marcy, Randolph B., 1853

Gypsum, Oklahoma
Van Vleet, A. H., 1904

Honduras
Moe, Alfred K., 1904

Idaho
Ashburner, William, 1869
Browne, J. R., 1867
Browne, J. R., 1868
Mullan, John, 1863
Peale, A. C., 1872
Raymond, Rossiter W., 1869
Raymond, Rossiter W., 1870
Raymond, Rossiter W., 1871
Raymond, Rossiter W., 1872
Raymond, Rossiter W., 1873
Raymond, Rossiter W., 1874
Raymond, Rossiter W., 1875
Raymond, Rossiter W., 1877

Iron
Leith, Charles K., 1907

Iron, Alabama
Alabama, University of, 1902

Iron, Arkansas
Linn, L. F., & Sevier, A. H., 1836

Iron, Illinois, Iowa
Locke, John, 1844
Owen, David D., 1840

Iron, Lake Superior region
Foster, J. W., & Whitney, J. D., 1851
Jackson, Charles T., 1849

Iron, Louisiana
Johnson, Lawrence C., 1888

Iron, Michigan, Minnesota
Foster, J. W., & Whitney, J. D., 1851

Iron, North Carolina
Laidley, T. T. S., 1856
Wilkes, Charles, Hunt, H., & Martin, D. B., 1859

Iron, Oregon
Gabb, W. M., 1867

Iron, Texas
Johnson, Lawrence C., 1888

Iron, United States
Pumpelly, Raphael, 1886

Iron, Wisconsin
Foster, J. W., & Whitney, J. D., 1851
Locke, John, 1844
Owen, David D., 1840

Kansas
Elliott, R. S., 1872

Lake Superior region
Featherstonhaugh, George W., 1836
Foster, J. W., & Whitney, J. D., 1850a

Lead, England
Bell, William H., 1844

Lead, Illinois, Iowa, Wisconsin
Locke, John, 1844
Owen, David D., 1840

Lead, Wisconsin
Bell, William H., 1844

Limestone
Eckel, Edwin C., 1903

Economic geology (Cont'd.)

Limestone, Alabama
Eckel, Edwin C., & Crider, A. F., 1905
Smith, Eugene A., 1903

Limestone, California
Williamson, R. S., 1850

Limestone, Mississippi
Eckel, Edwin C., & Crider, A. F., 1905

Limestone, Mississippi Valley
U. S. Department of Ariculture, 1870

Limestone, Oregon
Frazer, John F., 1850

Limestone, United States, eastern
U. S. Department of Agriculture, 1869c

Limestone, United States, southern
U. S. Department of Agriculture, 1869a

Limestone, Virginia
U. S. Department of Agriculture, 1869b

Mercury, Bolivia, Peru
Gibbon, Lardner, 1854

Metals, U. S. S. R.
Palmer, Aaron H., 1848

Mexico
Bureau of the American Republics, 1904b

Michigan
Foster, J. W., & Whitney, J. D., 1850b

Minnesota
Taylor, James W., 1867
Taylor, James W., 1868

Montana
Browne, J. R., 1867
Browne, J. R., 1868
Eaton, A. K.; Keyes, W. S., & DeLacy, W. W., 1869
Keyes, W. S., 1868
Mullan, John, 1863
Peale, A. C., 1872
Raymond, Rossiter W., 1869
Raymond, Rossiter W., 1870
Raymond, Rossiter W., 1871
Raymond, Rossiter W., 1872
Raymond, Rossiter W., 1873
Raymond, Rossiter W., 1874
Raymond, Rossiter W., 1875
Raymond, Rossiter W., 1877
Taylor, James W., 1867
Taylor, James W., 1868
U. S. Department of Agriculture, 1872

Nevada
Browne, J. R., 1867
Browne, J. R., 1868
Raymond, Rossiter W., 1869
Raymond, Rossiter W., 1870
Raymond, Rossiter W., 1871
Raymond, Rossiter W., 1872
Raymond, Rossiter W., 1873
Raymond, Rossiter W., 1874

Raymond, Rossiter W., 1875
Raymond, Rossiter W., 1877
New Hampshire
Taylor, James W., 1867
New Mexico
Bureau of immigration, 1901
Herrick, C. L., 1900
McCauley, C. A. H., 1878
Raymond, Rossiter W., 1870
Raymond, Rossiter W., 1871
Raymond, Rossiter W., 1872
Raymond, Rossiter W., 1873
Raymond, Rossiter W., 1874
Raymond, Rossiter W., 1877
Taylor, James W., 1867
Taylor, James W., 1868
North Carolina
Taylor, James W., 1867
North Dakota
Raymond, Rossiter W., 1877
Oklahoma
Van Vleet, A. H., 1907
Oregon
Browne, J. R., 1867
Browne, J. R., 1868
Raymond, Rossiter W., 1870
Raymond, Rossiter W., 1871
Raymond, Rossiter W., 1872
Raymond, Rossiter W., 1873

Raymond, Rossiter W., 1874
Raymond, Rossiter W., 1875
Raymond, Rossiter W., 1877
Panama
Blanchard, I., 1869
Raymond, Rossiter W., 1869
Petroleum
Peckham, S. F., 1884
Petroleum, Dominican Republic
Siguel, F., 1871
Petroleum, New Mexico
Herrick, C. L., 1901
Petroleum, United States
Hays, S. S., 1866
Philippines
Becker, George F., 1901
Eveland, Arthur J., 1907
McCaskey, H. D., 1905a
McCaskey, H. D., 1905b
McCaskey, H. D., 1907
Phosphate, South Carolina
Shaler, N. S., 1873
Phosphate, United States, southern
U. S. Department of Agriculture, 1869a
Platinum
Blake, William P., 1869
Platinum, Pacific slope
Walcott, Charles D., 1906
Rocky Mountains
Dodge, G. M., 1868
Dodge, G. M., 1869
Mullan, John, 1863
Peale, A. C., 1872

Economic geology (Cont'd.)

Raymond, Rossiter W., 1869
Raymond, Rossiter W., 1870
Raymond, Rossiter W., 1871
Raymond, Rossiter W., 1872
Raymond, Rossiter W., 1873
Raymond, Rossiter W., 1874
Raymond, Rossiter W., 1875
Raymond, Rossiter W., 1877
Taylor, James W., 1867
Taylor, James W., 1868
Van Lennep, D., 1868

Salt, Brazil
Herndon, Lewis, 1853

Salt, Colombia
Bureau of the American Republics, 1905a

Salt, Oklahoma
Van Vleet, A. H., 1904

Silver
Blake, William P., 1869

Silver, Andes Mountains, Bolivia, Peru
Gibbon, Lardner, 1854

Silver, Brazil
Herndon, Lewis, 1853

Silver, Lake Superior region
Jackson, Charles T., 1849

Silver, Nevada
Blatchly, A., 1867
Edmunds, J. M., et al, 1863

Silver, United States
King, Clarence; Emmons, S. F., & Becker, G. F., 1885

South Carolina
Taylor, James W., 1867

South Dakota
Jenny, Walter P., 1876
Raymond, Rossiter W., 1877
Taylor, James W., 1867
Taylor, James W., 1868

Tin, England
Hughes, George W., 1844

Trinidad
Wall, G. P., & Sawkins, Joseph, 1857

United States
Featherstonhaugh, George W., 1833
U. S. Department of the Interior, 1866

United States, central
Taylor, James W., 1867
Taylor, James W., 1868

United States, eastern
Featherstonhaugh, George W., 1836
Taylor, James W., 1867
Taylor, James W., 1868

United States, western
Browne, J. R., 1867
Browne, J. R., 1868

Utah
Browne, J. R., 1867
Browne, J. R., 1868
Peale, A. C., 1872
Raymond, Rossiter W., 1869
Raymond, Rossiter W., 1870

Raymond, Rossiter W., 1871
Raymond, Rossiter W., 1872
Raymond, Rossiter W., 1873
Raymond, Rossiter W., 1874
Raymond, Rossiter W., 1875
Raymond, Rossiter W., 1877
Taylor, James W., 1867
Taylor, James W., 1868

Venezuela
Goiticoa, N. V., 1904

Virginia
Taylor, James W., 1867

Washington
Browne, J. R., 1867
Browne, J. R., 1868
Mullan, John, 1863

Wisconsin
Featherstonhaugh, George W., 1836

Wyoming
Darton, Nelson H., 1906
Peale, A. C., 1872
Raymond, Rossiter W., 1870
Raymond, Rossiter W., 1871
Raymond, Rossiter W., 1872
Raymond, Rossiter W., 1873
Raymond, Rossiter W., 1877

Zinc, Iowa, Wisconsin
Locke, John, 1844
Owen, David D., 1840

Education
Economic geology
Curricula, Methods
Lyman, Benjamin S., 1869
Raymond, Rossiter W., 1869
Ward, Willard P., 1869
Geological exploration
Methods
Meek, F. B., 1872a
Geology
Lapworth, Charles, 1904
Geomorphology
Davis, W. M., 1905b
Marbut, C. F., 1905
Glacial geology
Methods
Agassiz, Louis, 1872a
Mineralogy
Roessler, A. R.; Wilson, Joseph, S., & Goodwin, J. R., 1871
Hughes, George W., et al, 1837

Egypt
Paleontology
Vertebrata, Tertiary
Andrews, C. W., 1907

Elastic properties
Earth
Chree, C., 1893

Elasticity. See Elastic properties.

Electricity
Earth
Matteucci, Carlo, 1871

Engineering geology

Bridges

Mississippi Valley
Warren, G. K., 1878

Canals

Central America, Colombia, Honduras, Mexico, Nicaragua, Panama
Davis, C. H., 1866

Colombia
Michler, N., 1861

Colombia, Panama
Collins, Frederick, 1879
Selfridge, Thomas O., 1871
Selfridge, Thomas O., 1874

Index of publications
U. S. Army Corps of Engineers, 1903

Mexico, Oaxaca, Vera Cruz
Fuertes, E. A., 1872
Shufeldt, Robert W., 1872

Mississippi delta
U. S. Army Corps of Engineers, 1874

Nicaragua
Dutton, C. E., 1902
Hatfield, Chester, 1874
Wheeler, E. S., 1902

Panama
Burr, William H., 1903
Davis, George W., et al, 1906
Lull, E. P., 1879
Nicaragua Canal Commission, 1906

Drainage changes

Illinois
Long, S. H.; Graham, R., & Philips, Joseph, 1819

Mississippi delta
Graham, George, 1829

Floods

Index of publications
U. S. Army Corps of Engineers, 1903

Mississippi delta
Ellet, Charles, Jr., 1852

Mississippi Valley
Cowden, John, 1878
Warren, G. K., 1875a

Index of publications
U. S. Army Corps of Engineers, 1903

Materials, properties
Merrill, George P., 1889

Railroads

Alaska, Alberta, British Columbia
Powell, John W., 1886b

Arizona
Parke, John G., 1855a

California
Abbot, Henry L., 1855
Williamson, R. S., 1855

California, Great Basin, Great Plains, Rocky Mountains
Beckwith, E. G., 1855a
Parke, John G., 1855a
Schiel, James, 1855a
Whipple, A. W., 1855

Central America, Colombia, Honduras, Mexico, Nicaragua, Panama
Davis, C. H., 1866

Index

Great Plains, Rocky Mountains
Dodge, G. M., 1868
Dodge, G. M., 1869
Lander, Frederick W., 1855
Mullan, John, 1863
Stevens, I. I., 1855a
Stevens, I. I., 1855b
Van Lennep, D., 1868

Idaho, Montana, Washington
Mullan, John, 1863

Index of publications
U. S. Army Corps of Engineers, 1903

Louisiana, Texas
Pope, John, 1855

Montana, Yukon
Powell, John W., 1886b

New Mexico
Parke, John G., 1855a
Parke, John G., 1855b

Oregon, Washington
Stevens, I. I., 1855a
Stevens, I. I., 1855b

United States, western
Davis, Jefferson, 1855
Humphreys, A. A., &
 Warren, G. K., 1855
U. S. War Department, 1855a
U. S. War Department, 1855b
Whipple, A. W., 1855

Washington
Lander, Frederick W., 1855

Rivers. See also **Drainage changes** under **Engineering geology.**

California
Mendell, G. H., 1882

Index of publications
U. S. Army Corps of Engineers, 1903

Louisiana
Forshey, C. G., 1875

Minnesota
Warren, G. K., 1875b

Mississippi Valley
Warren, G. K., 1875a

Wisconsin
Warren, G. K., 1876

Roads

Arizona, Mexico, Sonora
Fergusson, D., 1863

Great Plains, Idaho
Mullan, John, 1863

Illinois, Indiana, Ohio
Macomb, Alexander; Knight, Jonathan, & Wevers, C. W., 1826

Index of publications
U. S. Army Corps of Engineers, 1903

Montana, Rocky Mountains, Washington
Mullan, John, 1863

United States, eastern
Howard, William, 1828
Long, S. H., 1827

Rock mechanics
Hallock, William, 1893

Shorelines

Massachusetts
Van Ingen, H. S., 1874

North Carolina
U. S. Army Corps of Engineers, 1905

Engineering geology (Cont'd.)

Soils
Chemistry
Cameron, Frank K., 1901
Physical properties
Briggs, Lyman J., 1901

Tunnels
Nevada
Wright, H. G.; Foster, J. G., & Newcomb, Wesley, 1872

Wells
Texas
Pope, John, 1858

Waste disposal
California
Mendell, G. H., 1882

England

Economic geology
Copper and tin, Cornwall and Devon
Hughes, George W., 1844
Lead, Copper
Bell, William H., 1844
Mining law
Raymond, Rossiter W., 1869
Stratigraphy
Hall, James, 1851c
Surveys
Geological and topographical surveys
Comstock, C. B., 1876

Epeirogenesis

Blytt, A., 1890

Erosion. See also **Shorelines; Shore features** under **Marine geology.**
Glaciers
Quaternary
Wallace, A. R., 1894

Estuary. See also **Coastal features** under **Geomorphology.**
Delaware River
Mitchell, Henry, 1884

Eulogy. See **Memoirs.**

Europe. See also names of individual countries.
Paleoclimatology
Quaternary
Geikie, James, 1899
Paleontology
Man, fossil
Obermaier, Hugues, 1906
Suess, Edward, 1873
Vertebrata, Mammalia
Rütimeyer, L., 1862

Evolution

Sollas, W. J., 1901
Fresh water fauna
Origin
Gill, Theodore N., 1905
Man, fossil
Foot
Anthony, Raoul, 1904
Processes
Cambrian
Brooks, William K., 1896

Expeditions. See **Surveys.**

Exploration. See **Geological exploration**; see also **Exploration** under **Rivers.**

F

Faults. See also **Deformation.**

Lineation
Hobbs, William H., 1905

Finland
Surveys
Land and marine surveys
Wheeler, George M., 1885

Floods. See under **Engineering geology**; **Fluvial features** under **Geomorphology.**

Florida
Areal geology
Agassiz, Louis, 1852
Smith, Eugene A., 1884c
Geomorphology
Keys
Agassiz, Louis, 1852
Landform description
Smith, Eugene A., 1884c
Lacustrine & paludal features
MacGonigle, John N., 1905
Shore features
Gibbs, Oliver W., 1856
Soils
Smith, Eugene A., 1884c

Soils, Escambia County
Griffen, A. M., et al, 1908
Soils, Gadsden County
Fippin, Elmer O., & Root, Aldert S., 1904
Soils, Gainesville
Rice, Thomas D., & Geib, W. J., 1905b
Soils, Leon County
Wilder, Henry J., et al, 1907b
Marine geology
Reefs
Pourtales, L. F., 1871
Reefs, Sediments
Agassiz, Louis, 1852
Hunt, E. B., 1864
Shore features
Gibbs, Oliver W., 1856

Fluorspar
Greenland
Quale, Paul, 1867

Foraminifera
Atlantic Ocean
Pourtales, L. F., 1859
Louisiana, Recent
Hilgard, Eugene W., & Hopkins, F. V., 1878

Fractures. See also **Deformation.**
Faults, Joints
Lineation
Hobbs, William H., 1905

France
  Economic geology
    Education, Mining law
      Raymond, Rossiter W., 1869
  Paleontology
    Man, fossil
      Gaudry, Albert, 1903
  Surveys
    Land and marine surveys
      Wheeler, George M., 1885

G

Gas, natural
  Indiana, Ohio
    Orton, Edward, 1893

Gems
  Babinet, Jacques, 1871
  Brazil
    Bureau of the American Republics, 1905b
  Collections
    U. S. National Museum
      Kunz, George F., 1889
      Tassin, Wirt, 1902a

Geochemistry
  Hunt, T. Sterry, 1871

Geochronology. See also **Age** under **Earth.**
  Earth cooling
    Chamberlain, T. C., 1901
    Kelvin, William T., 1898
  Fossils
    White, Charles A., 1893
  History
    Geikie, Archibald, 1893
    Walcott, Charles D., 1894
  Ocean, dissolved salt
    Joly, J., 1901
  Rate of cooling
    King, Clarence, 1894
  Sedimentary rocks
    Walcott, Charles D., 1894
  Uniformitarianism
    Chamberlain, T. C., 1901
    Geikie, Archibald, 1893

Geodesy. See also **Earth.**
  Bibliography
    Gore, James H., 1903
  Figure of the earth
    Chree, C., 1893
  History
    Merino, Miguel, 1864
  1903
    Holdich, T. H., 1903

Geography
  Relation to geology
    Lapworth, Charles, 1904

Geological exploration. See also **Mineral exploration.**
  Alaska
    Allen, Henry T., 1887
    Barnard, E. C., 1899
    Blake, Theodore A., 1869
    Davidson, George, 1868
    Eldridge, G. H., & Muldrow, Robert, 1899
    Mendenhall, Walter C., 1899

Peters, W. J., & Brooks, Alfred H., 1899
Raymond, Charles W., 1871
Senate Committee on Military Affairs, 1900
Schrader, Frank C., 1899
Spurr, J. E., & Post, W. S., 1899
Walcott, Charles D., 1897

Andes Mountains

Gibbon, Lardner, 1854
Macrae, Archibald, 1855

Antarctica

Nordenskiöld, Otto, et al, 1904

Arctic

Final report

Hall, Charles F., 1879

Argentina

Macrae, Archibald, 1855

Arizona

Adams, Samuel, 1870
Antisell, Thomas, 1855
Beale, Edward F., 1858
Beale, Edward F., & Engle, F. E., 1860
Emory, William H., 1855
Emory, William H., 1857
Fergusson, D., 1863
Gilbert, G. K., 1872
Ives, Joseph C., 1858
Ives, Joseph C., 1861
Lockwood, Daniel W., 1872
Lyle, D. A., 1872
Michler, N., 1857b
Newberry, J. S., 1861
Parke, John G., 1855a
Parry, C. C., 1869
Parry, C. C., & Schott, Arthur, 1857
Powell, John W., 1872a
Powell, John W., 1872b
Powell, John W., 1873

Powell, John W., 1877
Simpson, J. H., 1850b
Sitgreaves, L., 1853
Wheeler, George M., 1872a
Wheeler, George M., 1872b
Wheeler, George M., 1873
Wheeler, George M., 1874
Wheeler, George M., 1875
Wheeler, George M., 1876
Wheeler, George M., 1877
Wheeler, George M., 1878
Wheeler, George M., 1879a
Whipple, A. W., 1851

Arkansas

Beale, Edward F., & Engle, F. E., 1860
Featherstonhaugh, George W., 1835
Marcy, Randolph B., 1850
Simpson, J. H., 1850a
Simpson, J. H., & Marcy, R. B., 1850

Bolivia

Gibbon, Lardner, 1854

Brazil

Herndon, Lewis, 1853

British Columbia

Blake, Theodore A., 1869

California

Abbot, Henry L., 1855
Adams, Samuel, 1870
Antisell, Thomas, 1855
Beckwith, E. G., 1855a
Blake, William P., 1855a
Blake, William P., 1855b
Blake, William P., 1856b
Chandler, M. T. W., 1857
Conkling, A. R., 1877a
Conkling, A. R., 1878
Derby, George H., 1850
Derby, George H., 1852b
Emory, William H., 1848
Emory, William H., 1857

Geological exploration (Cont'd.)

Fremont, John C., 1845
Fremont, John C., 1848
Gilbert, G. K., 1872
Ives, Joseph C., 1858
Ives, Joseph C., 1861
Johnston, A. R., 1858
King, Clarence, 1873
King, Clarence, 1874
King, Clarence, 1876
King, Clarence, 1877
Lyle, D. A., 1872
Marcou, Jules, 1855a
Marcou, Jules, 1855c
Marcou, Jules, 1876
Michler, N., 1857b
Newberry, J. S., 1855
Newberry, J. S., 1861
Parke, John G., 1855a
Parry, C. C., & Schott, Arthur, 1857
Schiel, James, 1855a
Sitgreaves, L., 1853
Wheeler, George M., 1874
Wheeler, George M., 1875
Wheeler, George M., 1876
Wheeler, George M., 1877
Wheeler, George M., 1878
Wheeler, George M., 1879a
Whipple, A. W., 1851
Whipple, A. W., 1855
Williamson, R. S., 1855

Chile

Gilliss, J. M., 1855
Macrae, Archibald, 1855

Colombia

Maack, G. A., 1874
Michler, N., 1861
Schott, Arthur, 1861b
Selfridge, Thomas O., 1874

Colorado

Adams, Samuel, 1870
Campbell, Donald, 1874
Conkling, A. R., 1877b
Hawn, F., 1874a
Hawn, F., 1874b
Hawn, R., 1874c
Hawn, Laurens, 1874
Hayden, F. B., 1872b
Hayden, F. V., 1877
Hayden, F. V., 1878b
Hayden, F. V., 1883a
King, Clarence, 1871
King, Clarence, 1873
King, Clarence, 1874
King, Clarence, 1876
King, Clarence, 1877
McCauley, C. A. H., 1878
Parry, C. C., 1869
Peale, A. C., 1873
Powell, John W., 1872a
Ruffner, E. H., 1874
Stevenson, J. T., 1879
Warren, G. K., 1858
Wheeler, George M., 1873
Wheeler, George M., 1874
Wheeler, George M., 1875
Wheeler, George M., 1876
Wheeler, George M., 1877
Wheeler, George M., 1878
Wheeler, George M., 1879a

Colorado Plateau

Geikie, Archibald, 1884
Ives, Joseph C., 1861
Newberry, J. S., 1861
Powell, John W., 1872a
Powell, John W., 1872b
Powell, John W., 1873

Dominican Republic

Adam, J. S., 1871
Blake, William P., 1871a
Blake, William P., 1871b

Great Basin

Beckwith, E. G., 1855a
Beckwith, E. G., 1855b
Blake, William P., 1855a
Emory, William H., 1848
Fremont, John C., 1848
Gilbert, G. K., 1872

Hall, James, 1852
Hayden, F. V., 1872a
Johnston, A. R., 1858
King, Clarence, 1871
King, Clarence, 1873
King, Clarence, 1874
King, Clarence, 1876
King, Clarence, 1877
Lockwood, Daniel W., 1872
Lyle, D. A., 1872
Marcou, Jules, 1855a
Marcou, Jules, 1855c
Schiel, James, 1855a
Stansbury, Howard, 1852
Wheeler, George M., 1872a
Wheeler, George M., 1872b
Wheeler, George M., 1873
Wheeler, George M., 1874
Wheeler, George M., 1875
Wheeler, George M., 1876
Wheeler, George M., 1877
Wheeler, George M., 1878
Wheeler, George M., 1879a
Whipple, A. W., 1855

Great Plains

Abert, J. W., 1846
Beckwith, E. G., 1855a
Beckwith, E. G., 1855b
Blake, William P., 1855a
Cross, Osborne, 1850
Emory, William H., 1848
Engelmann, Henry, 1858
Fremont, John C., 1843
Fremont, John C., 1845
Hall, James, 1852
Hayden, F. V., 1872b
Lander, Frederick W., 1855
Marcou, Jules, 1855a
Marcou, Jules, 1855c
Marcy, Randolph B., 1853
Mullan, John, 1863
Owen, David D., 1852
Schiel, James, 1855a
Shumard, B. F., 1853b
Stansbury, Howard, 1852
Stevens, I. I., 1855a

Stevens, I. I., 1855b
Warren, G. K., 1856
Warren, G. K., 1858
Whipple, A. W., 1855

History, see under Surveys

Idaho

Bradley, F. H., 1873
Hayden, F. V., 1872a
Hayden, F. V., 1873
Hayden, FV., 1877
Hayden, F. V., 1878b
Hayden, F. V., 1883a
Mullan, John, 1863
Wheeler, George M., 1877
Wheeler, George M., 1878
Wheeler, George M., 1879a

Illinois, Iowa

Locke, John, 1844
Nicollet, I. N., 1843
Owen, David D., 1840

Iowa

Allen, J., 1846

Israel

Dead Sea

Lynch, W. F., 1848

Japan

Jones, George, 1856b
Perry, M. C., 1856

Kansas

Engelmann, Henry, 1858
Hayden, F. V., 1858
Warren, G. K., 1858
Wheeler, George M., 1878
Wheeler, George M., 1879a

Lake Superior region

Allen, J., 1834
Featherstonhaugh, George W., 1836
Foster, J. W., & Whitney, J. D., 1850b
Jackson, Charles T., 1849
Pope, John, 1850

Geological exploration (Cont'd.)

Louisiana
Blake, William P., 1855c
Featherstonhaugh, George W., 1835
Marcou, Jules, 1855b
Pope, John, 1855

Manchuria
Collins, P. McD., 1858

Methods
Meek, F. B., 1872a

Mexico

Baja California
Chandler, M. T. W., 1857
Derby, George H., 1852a
Ives, Joseph C., 1858
Ives, Joseph C., 1861
Michler, N., 1857b
Newberry, J. S., 1861
Parry, C. C., & Schott, Arthur, 1857

Chihuahua, Coahuila
Michler, N., 1857a
Parry, C. C., & Schott Arthur, 1857
Wislezenus, A., 1848

Colorado River
Derby, George H., 1852a

Northern
Emory, William H., 1857

Nuevo Leon, Tamaulipas
Parry, C. C., & Schott, Arthur, 1857
Wislezenus, A., 1848

Oaxaca
Shufeldt, Robert W., 1872
Spear, John C., 1872

Sonora
Fergusson, D., 1863
Michler, N., 1857b

Vera Cruz
Parry, C. C., & Schott, Arthur, 1857

Shufeldt, Robert W., 1872
Spear, John C., 1872

Michigan
Foster, J. W., & Whitney, J. D., 1850b

Minnesota
Pope, John, 1850

Mississippi Valley, Missouri
Nicollet, I. N., 1843

Missouri
Featherstonhaugh, George W., 1835

Montana
Barlow, J. W., 1872
Doane, Gustavus C., 1871
Hayden, F. V., 1872a
Hayden, F. V., 1873
Hayden, F. V., 1878b
Mullan, John, 1863
Peale, A. C., 1873
Wheeler, George M., 1874
Wheeler, George M., 1875
Wheeler, George M., 1876
Wheeler, George M., 1877
Wheeler, George M., 1878
Wheeler, George M., 1879a

Nebraska
Engelmann, Henry, 1858
Hayden, F. V., 1858
Hayden, F. V., 1871
Hayden, F. V., 1872b
Owen, David D., 1852
Warren, G. K., 1856
Warren, G. K., 1858
Wheeler, George M., 1875
Wheeler, George M., 1878
Wheeler, George M., 1879a

Nevada

Conkling, A. R., 1877a
Conkling, A. R., 1878
Gilbert, G. K., 1872
Ives, Joseph C., 1861
King, Clarence, 1871
King, Clarence, 1873
King, Clarence, 1874
King, Clarence, 1876
King, Clarence, 1877
Lockwood, Daniel W., 1872
Lyle, D. A., 1872
Newberry, J. S., 1861
Powell, John W., 1877
Sitgreaves, L., 1853
Wheeler, George M., 1872a
Wheeler, George M., 1872b
Wheeler, George M., 1873
Wheeler, George M., 1874
Wheeler, George M., 1875
Wheeler, George M., 1876
Wheeler, George M., 1877
Wheeler, George M., 1878
Wheeler, George M., 1879a

New Mexico

Abert, J. W., 1846
Abert, J. W., 1848
Antisell, Thomas, 1855
Beale, Edward F., 1858
Beale, Edward F., & Engle, F. E., 1860
Conkling, A. R., 1876
Conkling, A. R., 1877b
Emory, William H., 1855
Emory, William H., 1857
Ives, Joseph C., 1858
Johnston, A. R., 1858
McCauley, C. A. H., 1878
Marcy, Randolph B., 1850
Parke, John G., 1855a
Parke, John G., 1855b
Parry, C. C., & Schott, Arthur, 1857
Ruffner, E. H., 1874
Simpson, J. H., 1850a
Simpson, J. H., 1850b

Simpson, J. H., & Marcy, R. B., 1850
Sitgreaves, L., 1853
Stevenson, J. T., 1879
Wheeler, George M., 1873
Wheeler, George M., 1874
Wheeler, George M., 1875
Wheeler, George M., 1876
Wheeler, George M., 1877
Wheeler, George M., 1878
Wheeler, George M., 1879a
Wislezenus, A., 1848

Nicaragua

Hatfield, Chester, 1874
Whitfield, Benjamin, 1874

Oklahoma

Marcy, Randolph B., 1853
Ruffner, E. H., 1877a
Shumard, B. F., 1853b
Wheelock, T. B., 1834

Oregon

Abbot, Henry L., 1855
Cross, Osborne, 1850
Fremont, John C., 1845
Fremont, John C., 1848
Newberry, J. S., 1855
Stevens, I. I., 1855a
Stevens, I. I., 1855b
Wheeler, George M., 1877
Wheeler, George M., 1878
Wheeler, George M., 1879a

Pacific Islands

Fah, C. F., 1856
Perry, M. C., 1856
Taylor, Bayard, 1856

Panama

Bowditch, E. W., 1874
Carson, J. Petigru, 1874
Engle, F., 1860
Maack, G. A., 1874
Selfridge, Thomas O., 1874

Geological exploration (Cont'd.)

Peru
Gibbon, Lardner, 1854

Philippines
McCaskey, H. D., 1907

Rocky Mountains
Abert, J. W., 1846
Adams, Samuel, 1870
Beckwith, E. G., 1855a
Beckwith, E. G., 1855b
Blake, William P., 1855a
Cross, Osborne, 1850
Doane, Gustavus C., 1871
Emory, William H., 1848
Fremont, John C., 1843
Fremont, John C., 1845
Hall, James, 1852
Hayden, F. V., 1872a
Hayden, F. V., 1872b
Hayden, F. V., 1877
Lander, Frederick W., 1855
Marcou, Jules, 1855a
Marcou, Jules, 1855c
Mullan, John, 1863
Powell, John W., 1877
Schiel, James, 1855a
Stansbury, Howard, 1852
Stevens, I. I., 1855a
Stevens, I. I., 1855b
Wheeler, George M., 1874
Wheeler, George M., 1875
Wheeler, George M., 1876
Wheeler, George M., 1877
Wheeler, George M., 1878
Wheeler, George M., 1879a
Whipple, A. W., 1855

Samoa
Steinberger, A. B., 1874

South Dakota
Culbertson, Thaddeus A., 1851
Engelmann, Henry, 1858
Hayden, F. V., 1872b
Hayden, F. V., 1878b
Jenny, Walter P., 1875
Jenny, Walter P., 1876
Ludlow, William, 1875
Owen, David D., 1852
Powell, John W., 1877
Warren, G. K., 1856
Warren, G. K., 1858
Winchell, N. H., 1874
Winchell, N. H., 1875

Sudan
Chevalier, Auguste, 1905

Taiwan
Perry, M. C., 1856

Texas
Beale, Edward F., 1858
Beale, Edward F., & Engle, F. E., 1860
Blake, William P., 1855c
Blake, William P., 1856a
Emory, William H., 1857
Johnston, J. E., et al, 1850
Marcou, Jules, 1855b
Marcy, Randolph B., 1850
Marcy, Randolph B., 1853
Marcy, Randolph B., 1856
Michler, N., 1857a
Parry, C. C., & Schott, Arthur, 1857
Pope, John, 1855
Ruffner, E. H., 1877a
Ruffner, E. H., 1877b
Shumard, B. F., 1853b
Simpson, J. H., 1850a
Simpson, J. H., & Marcy, R. B., 1850
Wheeler, George M., 1877
Wheeler, George M., 1878
Wheeler, George M., 1879a
Wheelock, T. B., 1834
Whiting, William H. C., 1850
Wislezenus, A., 1848

Union of Soviet Socalist Republics

Collins, P. McD., 1858

United States See also **Geological exploration** under the names of states.

Featherstonhaugh, George W., 1833

Eastern

Featherstonhaugh, George W., 1836

Southwest

Emory, William H., 1857
Parry, C. C., & Schott, Arthur, 1857

Western

Blake, William P., 1855a
Davis, Jefferson, 1855
Humphreys, A. A., & Warren, G. K., 1855
Marcou, Jules, 1855a
Marcou, Jules, 1855c
U. S. War Department, 1855a
U. S. War Department, 1855b
U. S. War Department, 1874
Whipple, A. W., 1855

U. S. Army Engineers

Index of publications

U. S. Army Corps of Engineers, 1903

Utah

Adams, Samuel, 1870
Bradley, F. H., 1873
Dana, James D., 1882
Engelmann, Henry, 1859
Geikie, Archibald, 1882
Hall, James, 1852
Hayden, F. V., 1872a
Hayden, F. V., 1872b

Hayden, F. V., 1873
Hayden, F. V., 1877
Hayden, F. V., 1883a
Jones, William A., 1872
King, Clarence, 1871
King, Clarence, 1873
King, Clarence, 1874
King, Clarence, 1876
King, Clarence, 1877
Parry, C. C., 1869
Peale, A. C., 1873
Powell, John W., 1872a
Powell, John W., 1872b
Powell, John W., 1873
Powell, John W., 1877
Simpson, J. H., 1859
Stansbury, Howard, 1852
Wheeler, George M., 1872b
Wheeler, George M., 1873
Wheeler, George M., 1874
Wheeler, George M., 1875
Wheeler, George M., 1876
Wheeler, George M., 1877
Wheeler, George M., 1878
Wheeler, George M., 1879a

Washington

Gibbs, George, 1855
Lander, Frederick W., 1855
Mullan, John, 1863
Stevens, I. I., 1855a
Stevens, I. I., 1855b
Wheeler, George M., 1878
Wheeler, George M., 1879a

Wisconsin

Featherstonhaugh, George W., 1836
Locke, John, 1844
Nicollet, I. N., 1843
Owen, David D., 1840

Wyoming

Bannister, H. M., 1873
Barlow, J. W., 1872
Bradley, F. H., 1873
Comstock, Theodore B., 1874

Geological exploration (*Cont'd.*)
Doane, Gustavus C., 1871
Engelmann, Henry, 1858
Engelmann, Henry, 1859
Hayden, F. V., 1872a
Hayden, F. V., 1872b
Hayden, F. V., 1873
Hayden, F. V., 1877
Hayden, F. V., 1878b
Hayden, F. V., 1883a
Hayden, F. V., 1883b
Holmes, William H., 1883
Jones, William A., 1874
Peale, A. C., 1873
St. John, Orestes H., 1883
Simpson, J. H., 1859
Warren, G. K., 1856
Warren, G. K., 1858
Wheeler, George M., 1874
Wheeler, George M., 1875
Wheeler, George M., 1876
Wheeler, George M., 1877
Wheeler, George M., 1878
Wheeler, George M., 1879a

Geological surveys. See Surveys.

Geology, history of
Anderson, Martin B., 1871
Arago, D. F. J., 1872
Cuvier, Georges, 1861
de Quatrefages, J. L. A., 1863
Elie de Beaumont, L., 1871
Favre, Ernest, 1879
Fischer, P., 1873
Flourens, P., 1862
Flourens, P., 1863
Flourens, P., 1866
Flourens, P., 1869a
Flourens, P., 1869b
Gilbert, G. K., 1903
LeConte, Joseph, 1901
Prestwich, Joseph, 1876
Schuchert, Charles, 1905
Stebbins, Rufus P., 1874

1881
Hawes, George W., 1881a
1882
Hunt, T. Sterry, 1884
1883
Hunt, T. Sterry, 1885
1886
Darton, Nelson H., 1889
1887
McGee, W. J., 1890
1888
McGee, W. J., 1890

Geology and history
Suess, Edward, 1873

Memoirs
Agassiz, Louis
Stebbins, Rufus P., 1874
Bravais, A.
Elie de Beaumont, L., 1871
Cuvier, G.
Flourens, P., 1869b
de Blainville, D.
Flourens, P., 1866
Dewey, C.
Anderson, Martin B., 1871
Fourier, J.
Arago, D. F. J., 1872
Hauy, R. J.
Cuvier, Georges, 1861
Lartet, Edward
Fischer, P., 1873
Powell, J. W.
Gilbert, G. K., 1903

Saint Hilaire, G.
de Quatrefages, J. L. A.,
  1863
Flourens, P., 1862
von Buch, L.
Flourens, P., 1863
von Zittel, K. A.
Schuchert, Charles, 1905
United States
Merrill, George P., 1906

Geology, Methods
Meek, F. B., 1872a

Geology, relation to other sciences
Lapworth, Charles, 1904

Geomorphology
  Coastal features. See also Shorelines; Shore features under Marine geology.
  Agassiz, Louis, 1868
  Delaware
  Mitchell, Henry, 1884
  Florida
  Agassiz, Louis, 1852
  Gibbs, Oliver W., 1852
  Maine
  Mitchell, Henry, 1880
  Massachusetts
  Boutelle, Charles O., 1887
  Gulliver, F. P., 1905
  Marinden, Henry L., 1890a
  Marinden, Henry L., 1890b
  Marinden, Henry L., 1892a
  Marinden, Henry L., 1892b
  Marinden, Henry L., 1894
  Marinden, Henry L., 1896
  Marinden, Henry L., 1897

Mitchell, Henry, 1861
Mitchell, Henry, 1875
Mitchell, Henry, 1880
Mitchell, Henry, 1887a
Mitchell, Henry, 1889b
Van Ingen, H. S., 1874
Whiting, Henry L., 1875
Whiting, Henry L., 1887
Whiting, Henry L., 1890
  New Jersey
  Mitchell, Henry, 1884
  New York
  Bache, A. D., 1856b
  Nicaragua
  Mitchell, Henry, 1877
  North Carolina
  U. S. Army Corps of Engineers, 1905
  Nova Scotia
  Mitchell, Henry, 1880
  Oregon
  Langfitt, W. C., 1903
  Panama
  Mitchell, Henry, 1877
  Sediments, Chesapeake Bay
  Hughes, George W., 1837
  Tidal features
  Hilgard, J. E., 1875
  Washington
  Langfitt, W. C., 1903

Deltas
  Delaware River
  Mitchell, Henry, 1887b
  Mississippi Delta
  Delafield, Richard, 1829
  Ellet, Charles, Jr., 1852
  Forshey, C. G., 1875
  Graham, George, 1829
  Hilgard, Eugene W., & Hopkins, F. V., 1878

Geomorphology (Cont'd.)

Long, S. H., & Humphreys, A. A., 1851
U. S. Army Corps of Engineers, 1874

Education
Davis, W. M., 1905b
Marbut, C. F., 1905

Fluvial features
Gilbert, G. K., 1885

Alps Mountains
Penck, Albrecht, 1905

Arkansas
Linn, L. F., & Sevier, A. H., 1836

California
Mendell, G. H., 1882

Connecticut
Powell, Charles F., 1905

Delaware River
Marinden, Henry L., 1882
Marinden, Henry L., 1883
Marinden, Henry L., 1885
Mitchell, Henry, 1881
Mitchell, Henry, 1884
Mitchell, Henry, 1889a

Drainage changes, Illinois
Long, S. H.; Graham, R., & Philips, Joseph, 1819

Floodplains
Long, S. H., & Humphreys, A. A., 1851

Log dams
Fuller, Charles A., 1855
Linn, L. F., & Sevier, A. H., 1836
Linnard, T. B., 1844
Long, S. H., 1841
Paxton, Joseph, 1829
Shreve, Henry M., 1833

Shreve, Henry M., & Gratiot, C., 1834

Louisiana
Forshey, C. G., 1875

Massachusetts
Powell, Charles F., 1905

Meanders
Bernard, S., & Totten, J. G., 1823

Minnesota
Warren, G. K., 1875b

Mississippi River
Graham, George, 1829
Bernard, S., & Totten, J. G., 1823
Little, George, 1882
Mitchell, Henry, 1883
U. S. Army Corps of Engineers, 1874

Mississippi Valley
Cowden, John, 1878
Warren, G. K., 1875a
Warren, G. K., 1878

New Jersey
Mitchell, Henry, 1881
Mitchell, Henry, 1884
Mitchell, Henry, 1889a

New York
Tarr, R. S., 1905

Niagara River
Gilbert, G. K., 1891

Nicaragua
Davis, Arthur P., 1903

Ohio River
Bernard, S., & Totten, J. G., 1823

Oregon
Langfitt, W. C., 1903

Panama
Davis, Arthur P., 1903

Patterns
Gilbert, G. K., 1885

Pennsylvania
Mitchell, Henry, 1881
Mitchell, Henry, 1884

San Joaquin River
Soulé, Frank, 1902

Sediments, Mississippi River
Hildgard, Eugene W., &
Hopkins, F. V., 1878

Sediments, Susquehanna River
Hughes, George W., 1837

Tertiary Rivers, California
Bowman, Amos, 1873

Washington
Langfitt, W. C., 1903

Water gaps
Howard, William, 1828

Waterfalls, Niagara River
Gilbert, G. K., 1891

Wisconsin
Warren, G. K., 1876

Yuba River
Manson, Marsden, 1902

Glacial features
Geikie, James, 1891

Alaska, Muir glacier
Reid, Harry F., 1892

Alps Mountains
Penck, Albrecht, 1905

Antarctica
Arctowski, Henryk, 1902

Great Basin, Great Plains, Rocky Mountains
Newberry, J. S., 1872

Great Lakes region
Gilbert, G. K., 1899

Hungary
de Martonne, E., 1905

India, Kashmir
Workman, Fanny B., 1905
Workman, W. H., 1905

Iowa
Owen, David D., 1848b

Isostatic rebound
Bell, Robert, 1898

Lake Superior region
Desor, E., 1851

Lake Superior region, Minnesota
Owen, David D., 1848b

Montana
Chaney, L. W., 1905
Matthes, Francois E., 1905

New York
Tarr, R. S., 1905

New York, Ontario, Quebec
Colemen, A. P., 1905

Oregon
Reid, Harry F., 1905a

Peru
Pfordte, Otto F., 1905

Roumania
de Martonne, E., 1905

Washington
Reid, Harry F., 1905a

Wisconsin
Owen, David D., 1848b

Lacustrine features
Alps Mountains
Penck, Albrecht, 1905

Geomorphology (Cont'd.)

California, Alkaline lakes
Loew, Oscar, 1876a

Florida
MacGonigle, John N., 1905

Great Basin, Great Plains, Rocky Mountains
Newberry, J. S., 1872

Great Lakes region
Gilbert, G. K., 1899

Lake Superior region
Desor, E., 1851

Louisiana
Forshey, C. G., 1875

Michigan, Wisconsin
Bixby, W. H.; Beach, Lansing H., & Gaillard, D. D., 1906

Oregon, Crater Lake
Diller, J. S., 1898

Landform description

Alabama
Smith, Eugene A., 1884b

Alaska
Blake, Theodore, A., 1869
Brooks, Alfred H., 1905
Bryant, Charles, 1870
Bulkley, Charles S., 1868
Cook, Frederick A., 1905
Davidson, George, 1868
McIntyre, H. H., 1870
Raymond, Charles W., 1871
Reid, Harry F., 1892
Schwatka, Frederick, 1884

Antartica
Arctowski, Henryk, 1902

Arctic
Hall, Charles F., 1879

Argentina
Bureau of the American Republics, 1903

Arizona
Lacey, John F., 1906
Powell, John W., 1878b

Arkansas
Loughridge, R. H., 1884b

Australia
Thomson, J. P., 1898

Baja California, Baja California Sur
Merrill, George P., 1897

Bolivia
Bureau of the American Republics, 1904a

Brazil
Bureau of the American Republics, 1902

British Columbia
Blake, Theodore A., 1869

California
Hilgard, Eugene W., 1884b
Loew, Oscar, 1876b
Mendell, G. H., 1882
Powell, John W., 1878b
Soulé, Frank, 1902

Caribbean region
U. S. War Department, 1906a
U. S. War Department, 1906b

China, Tibet
Younghusband, Frank, 1906

Colombia
Collins, Frederick, 1879
Schott, Arthur, 1861b
Selfridge, Thomas O., 1874

Index

Colorado
McCauley, C. A. H., 1878
McCauley, C. A. H., 1879
Powell, John W., 1878b

Dominican Republic
McClellan, George B., 1871

Florida
Smith, Eugene A., 1884c

Georgia
Keith, Arthur, 1902
Smith, Eugene A., 1884a

Great Basin
Powell, John W., 1878b
Thomas, Cyrus, 1872
U. S. Department of Agriculture, 1871

Great Plains
Dodge, G. M., 1869
Powell, John W., 1878b
Thomas, Cyrus, 1872
Van Lennep, D., 1868

Honduras
Moe, Alfred K., 1904

Hungary
de Martonne, E., 1905

Idaho
Langford, N. P., 1873a

Israel
Lynch, W. F., 1848

Keewatin, Labrador, Mackenzie, Manitoba
Wilson, A. W. G., 1905

Labrador
Lieber, Oscar M., 1861

Lake Superior region, Michigan
Foster, J. W., 1849b

Louisiana
Hilgard, Eugene W., 1884c

Mexico
Bureau of the American Republics, 1904b
Hill, Robert T., 1905a

Minnesota
Hayden, F. V., 1873
Thomas, Cyrus, 1873
Warren, G. K., 1875b

Mississippi
Hilgard, Eugene W., 1884d

Mississippi Valley
Hilgard, Eugene W., 1884a
Little, George, 1882

Missouri
Loughridge, R. H., 1884a

Montana
Dana, Edward S., & Grinnell, George B., 1876
Doane, Gustavus C., 1871
Ludlow, William, 1876
Matthes, Francois E., 1905
Thomas, Cyrus, 1872
U. S. Department of Agriculture, 1872

Morocco
Fischer, Theobald, 1905

Nebraska
Hayden, F. V., 1873
Thomas, Cyrus, 1873

Nevada
Powell, John W., 1878b
U. S. Department of Agriculture, 1871

Newfoundland
Lieber, Oscar M., 1861

New Mexico
McCauley, C. A. H., 1878
Powell, John W., 1878b

Nicaragua
Hatfield, Chester, 1874
Whitfield, Benjamin, 1874

Geomorphology (Cont'd.)

North America
Froebel, Julius, 1855

North Carolina
Keith, Arthur, 1902
Kerr, W. C., 1884a

North Dakota
Hayden, F. V., 1873
Thomas, Cyrus, 1873

Northwest Territories
Hall, Charles F., 1879

Oklahoma
Loughridge, R. H., 1884d

Ontario, Quebec
Wilson, A. W. G., 1905

Oregon, Columbia River
Hergesheimer, E., 1883

Panama
Collins, Frederick, 1879
Lull, E. P., 1879
Selfridge, Thomas O., 1874

Philippines
Eveland, Arthur J., 1907

Rocky Mountains
Doane, Gustavus C., 1871
Dodge, G. M., 1869
Powell, John W., 1878b
Thomas, Cyrus, 1872
Van Lennep, D., 1868

Roumania
de Martonne, E., 1905

South Carolina
Hammond, Harry, 1884
Keith, Arthur, 1902

South Dakota
Hayden, F. V., 1873
Hoffman, William, 1879
Thomas, Cyrus, 1873

Tennessee
Keith, Arthur, 1902
Safford, James M., 1884

Texas
Loughridge, R. H., 1884c
Powell, John W., 1878b

United States
Whitney, J. D., 1874

Utah
Jones, William A., 1872
Powell, John W., 1878b
Thomas, Cyrus, 1872
U. S. Department of Agriculture, 1871

Venezuela
Goiticoa, N. V., 1904

Virginia
Keith, Arthur, 1902
Kerr, W. C., 1884b

Washington, Columbia River
Hergesheimer, E., 1883

West Indies, Isle of Pines
Hayes, Willard, 1906

Wyoming
Dana, Edward S., & Grinnell, George B., 1876
Doane, Gustavus C., 1871
Gannett, E. M., 1883
Hayden, F. V., 1883b
Holmes, William H., 1883
Langford, N. P., 1873b
Ludlow, William, 1876
St. John, Orestes H., 1883
Thomas, Cyrus, 1872

Landform evolution
Davis, W. M., 1904
Davis, W. M., 1905a

Mountains
Rice, William N., 1905

Index

Loess
Mississippi Valley
Little, George, 1882
Methods
Coastal features
Agassiz, Louis, 1868
Map drawing
deMello, Carlos, 1905
Paludal features
Florida
MacGonigle, John N., 1905
Soils
Alabama
Avon-Burke, R. T., et al, 1903
Avon-Burke, R. T., et al, 1904
Bennett, Frank, Jr., & Griffen, A. M., 1904
Bonsteel, F. E., et al, 1907
Carr, E. P., et al, 1907
Hearn, W. Edward, & Geib, W. J., 1908
Jones, Grove B., & Carr, M. E., 1904
McLendon, W. E., & Mann, Charles J., 1907
Smith, Eugene A., 1884b
Smith, William G., & Meeker, F. N., 1905
Smith, William G., & Meeker, F. N., 1907
Wilder, Henry J., & Bennett, Hugh H., 1905

Arizona
Holmes, J. Garnett, 1903b
Holmes, J. Garnett, et al, 1905b
Lapham, Macy H., & Neill, N. P., 1904
Means, Thomas H., 1901

Arkansas
Carter, William T., Jr., et al, 1908
Lapham, J. E., 1903
Loughridge, R. H., 1884b
Martin, J. O., & Carr, E. P., 1904b
Wilder, Henry J., & Shaw, Charles F., 1908

California
Hilgard, Eugene W., 1884b
Holmes, J. Garnett, 1901
Holmes, J. Garnett, 1902
Holmes, J. Garnett, et al, 1904a
Holmes, J. Garnett, et al, 1904b
Holmes, J. Garnett, et al, 1905a
Holmes, J. Garnett, et al, 1905b
Holmes, J. Garnett, & Mesmer, Louis, 1902
Lapham, Macy H., 1904
Lapham, Macy H., & Heileman, W. H., 1902a
Lapham, Macy H., & Heileman, W. H., 1902b
Lapham, Macy H., & Jensen, Charles A., 1905
Lapham, Macy H., & Mackie, W. W., 1905
Lapham, Macy H., & Mackie, W. W., 1907
Means, Thomas H., & Holmes, J. Garnett, 1901
Means, Thomas H., & Holmes, J. Garnett, 1902
Mesmer, Louis, 1904

Colorado
Holmes, J. Garnett, 1904
Holmes, J. Garnett, & Neill, N. P., 1905
Holmes, J. Garnett, & Rice, Thomas D., 1907

Geomorphology (Cont'd.)

Lapham, Macy H., et al, 1903
Means, Thomas H., 1900a

Connecticut
Dorsey, Clarence W., & Bonsteel, Jay A., 1900
Fippin, Elmer O., 1904

Delaware
Bonsteel, F. E., & Ayrs, O. L., 1904

Florida
Fippin, Elmer O., & Root, Aldert S., 1904
Griffen, A. M., et al, 1908
Rice, Thomas D., & Geib, W. J., 1905b
Smith, Eugene A., 1884c
Wilder, Henry J., et al, 1907b

Georgia
Avon-Burke, R. T., & Marean, Herbert W., 1902a
Carr, M. Earl, & Tharp, W. E., 1908
Ely, Charles W., & Griffen, A. M., 1905
Fippin, Elmer O., & Drake, J. A., 1905a
Lapham, J. E.; Lyman, W. S., & Ely, Charles, W., 1907
Marean, Herbert W., 1902
Smith, Eugene A., 1884a
Smith, William G., & Carter, William T., Jr., 1904a

Idaho
McLendon, W. E., 1904
Mesmer, Louis, 1903

Illinois
Bonsteel, Jay A., et al, 1903a
Bonsteel, Jay A., et al, 1903b
Coffey, George N., 1903b
Coffey, George N., et al, 1903a
Coffey, George N., et al, 1904a
Coffey, George N., et al, 1904b
Coffey, George N., et al, 1904c
Coffey, George N., et al, 1904d
Coffey, George N., et al, 1904e
Fippin, Elmer O., & Drake, J. A., 1905b

Indiana
Avon-Burke, R. T., & Ruhlen, La Mott, 1904a
Bennett, Frank, Jr., & Ely, Charles W., 1905
Mangum, A. W., & Neill, N. P., 1905a
Mangum, A. W., & Neill, N. P., 1905b
Marean, Herbert W., 1903a
Neill, N. P., & Tharp, W. E., 1907a
Neill, N. P., & Tharp, W. E., 1907b
Tharp, W. E., & Mann, Charles J., 1908

Iowa
Ely, Charles W.; Coffey, George N., & Griffen, A. M., 1905
Fippin, Elmer O., 1903a
Marean, Herbert W., & Jones, Grove B., 1904a
Marean, Herbert W., & Jones, Grove B., 1904b

Kansas

Burgess, James L., & Coffey, George N., 1905
Burgess, James L.; Tharp, W. E., & Lyman, W. S., 1907
Carter, William T., Jr., & Smith, Howard C., 1908
Drake, J. A., 1904
Drake, J. A., & Tharp, W. E., 1905
Lapham, J. E., & Olshausen, B. A., 1903
Mangum, A. W., & Drake, J. A., 1904

Kentucky

Avon-Burke, R. T., 1904a
Avon-Burke, R. T., 1904b
Griffen, A. M., & Ayrs, Orla L., 1907a
Marean, Herbert W., 1903b
Rice, Thomas D., 1907
Rice, Thomas D., & Geib, W. J., 1905d

Idaho

Jensen, Charles A., & Olshausen, B. A., 1902

Louisiana

Burgess, James L., et al, 1908
Ely, Charles W.; Marean, Herbert W., & Neil, N. P., 1907
Griffen, A. M., & Caine, Thomas A., 1907
Heileman, W. H., & Mesmer, Louis, 1902
Hilgard, Eugene W., 1884c
Jones, Grove B., & Ruhlen, La Mott, 1905
Rice, Thomas D., 1904
Rice, Thomas D., & Griswold, Lewis, 1904a
Rice, Thomas D., & Griswold, Lewis, 1904b

Maryland

Bonsteel, Jay A., 1901a
Bonsteel, Jay A., 1901b
Bonsteel, Jay A., et al, 1902a
Bonsteel, Jay A., & Avon-Burke, R. T., 1901
Bonsteel, F. E., & Carter, William T., Jr., 1904
Dorsey, Clarence W., & Bonsteel, Jay A., 1901
Smith, William G., & Martin, J. O., 1902

Massachusetts

Fippin, Elmer O., 1904

Michigan

Fippin, Elmer O., & Rice, Thomas D., 1902
Geib, W. J., 1908
Hearn, W. Edward, & Griffen, A. M., 1905
Jones, Grove B., & Carr, M. Earl, 1907
McLendon, W. E., & Carr, M. Earl, 1905
Mangum, A. W., & Mann, Charles J., 1905
Rice, Thomas D., & Geib, W. J., 1905c
Wilder, Henry J., & Geib, W. J., 1904b

Minnesota

Bennett, Hugh H., & Hurst, Lewis A., 1908
Caine, Thomas A., & Lyman, W. S., 1905b
Geib, W. J., & Jones, Grove B., 1907
Mangum, A. W., & Schroeder, F. C., 1908
Wilder, Henry J., 1904

Mississippi

Bennett, Frank, Jr., & Winston, R. A., 1908

Geomorphology (*Cont'd.*)

Bonsteel, Jay A., et al, 1902b
Burgess, James L., & Tharp, W. E., 1907
Caine, Thomas A., & Schroeder, Frank C., 1908
Hearn, W. Edward, & Carr, M. E., 1905
Hilgard, Eugene W., 1884d
Martin, J. O., & Ayrs, O. L., 1905
Smith, William G., & Carter, William T., Jr., 1903
Smith, William G., & Carter, William T., Jr., 1904b

Mississippi Valley
Hilgard, Eugene W., 1884a

Missouri
Avon-Burke, R. T., & Ruhlen, La Mott, 1904b
Carr, M. Earl, & Belden, H. L., 1905
Drake, J. A., & Strahorn, A. T., 1905
Fippin, Elmer O., 1903b
Fippin, Elmer O., & Drake, J. A., 1905b
Hearn, W. Edward, & Mann, Charles J., 1907a
Hearn, W. Edward, & Mann, Charles J., 1907b
Loughridge, R. H., 1884a
Mann, Charles J., & Tharp, W. E., 1908

Montana
Jensen, Charles A., & Neill, N. P., 1903b
Lapham, Macy H., & Ely, Charles W., 1907

Nebraska
Burgess, James L., & Worthen, H. L., 1908
Hearn, W. Edward, 1904
Hearn, W. Edward, & Burgess, James L., 1904a
Kocher, A. E., & Hurst, Lewis A., 1907b
Martin, J. O., & Sweet, A. T., 1905

New Hampshire
Mooney, Charles N.; Westover, H. L., & Bennett, Frank, 1908

New Jersey
Avon-Burke, R. T., & Wilder, Henry J., 1903
Bonsteel, Jay A., & Taylor, F. W., 1902

New Mexico
Means, Thomas H., & Gardner, Frank D., 1900

New York
Avon-Burke, R. T., & Marean, Herbert W., 1902b
Bonsteel, Jay A., et al, 1904
Bonsteel, F. E.; Carter, William T., Jr., & Ayrs, O. L., 1904
Bonsteel, Jay A.; Fippin, Elmer O., & Carter, William T., 1907
Carr, M. Earl; Griffin, A. M., & Lee, Ora, Jr., 1908
Fippin, Elmer O., & Carter, William T., 1907
Fippin, Elmer O., & Mann, C. W., 1908
Hearn, W. Edward, 1903
Lapham, J. E., & Bennett, Hugh H., 1905
Mesmer, Louis, & Hearn, W. E., 1903

Wilder, Henry J., & Belden, H. L., 1905b

North Carolina
Caine, Thomas A., 1903
Caine, Thomas A., & Mangum, A. W., 1903
Coffey, George N., & Hearn, W. Edward, 1902a
Coffey, George N., & Hearn, W. Edward, 1902b
Dorsey, Clarence W., et al, 1902
Drake, J. A., & Belden, H. L., 1908
Hearn, W. Edward, & MacNider, G. M., 1908a
Hearn, W. Edward, & MacNider, G. M., 1908b
Kerr, W. C., 1884a
Lapham, J. E., & Lyman, W. S., 1907
Lapham, J. E., & Meeker, F. N., 1904
Mooney, Charles N., & Ayrs, O. L., 1905a
Smith, William G., 1901
Smith, William G., & Coffey, George N., 1904
Root, Aldert S., & Hurst, Lewis A., 1907

North Dakota
Caine, Thomas A., 1904
Caine, Thomas A., & Kocher, A. E., 1904a
Ely, Charles W.; Willard, Rex E., & Weaver, J. T., 1908
Fippin, Elmer O., & Burgess, James L., 1905
Jensen, Charles A., & Neill, N. P., 1903a
Kocher, A. E., & Hurst, Lewis A., 1907a
Rice, Thomas D., 1908

Ohio
Caine, Thomas A., & Lyman, W. S., 1905c
Dorsey, Clarence W., & Coffey, George N., 1901
Lapham, J. E., & Mooney, Charles N., 1907a
Lapham, J. E., & Mooney, Charles N., 1907b
Martin, J. O., & Carr, E. P., 1904a
Meeker, F. N., & Tailby, G. W., 1908
Rice, Thomas D., & Geib, W. J., 1905a
Smith, William G., 1903a
Smith, William G., 1903b

Oklahoma
Loughridge, R. H., 1884d
McLendon, W. E., & Jones, Grove B., 1908
Rice, Thomas D., & Ayrs, Orla L., 1908

Oregon
Jensen, Charles A., 1904
Jensen, Charles A., & Mackie, W. W., 1904

Pennsylvania
Dorsey, Clarence W., 1901
Martin, J. O., 1904
Smith, William G., & Bennett, Frank, Jr., 1902
Wilder, Henry J., & Belden, H. L., 1905a
Wilder, Henry J., et al, 1907a
Wilder, Henry J.; Strahorn, A. T., & Geib, W. J., 1907

Philippines
Dorsey, Clarence W., 1907
Sanchez, Alfred M., 1905

Geomorphology (Cont'd.)

Porto Rico
Dorsey, Clarence W.; Mesmer, Louis, & Caine, Thomas A., 1903

Rhode Island
Bonsteel, F. E., & Carr, E. P., 1905b

South Carolina
Bennett, Frank, Jr., & Griffen, A. M., 1905
Bonsteel, F. E., & Carr, E. P., 1905a
Drake, J. A., & Belden, H. L., 1907a
Drake, J. A., & Belden, H. L., 1907b
Hammond, Harry, 1884
Mangum, A. W., & Root, Aldert S., 1904
Rice, Thomas D., & Taylor, F. W., 1903
Root, Aldert S., & Hurst, Lewis A., 1905
Taylor, F. W., 1903

South Dakota
Bennett, Frank, Jr., 1904

Tennessee
Carr, M. Earl, & Bennett, Frank, 1907
Lapham, J. E., & Miller, M. F., 1902
Lyman, W. S.; Bennett, Frank, & McLendon, W. E., 1908
McLendon, W. E., & Lyman, W. S., 1908
Mooney, Charles N., & Ayrs, O. L., 1905a
Mooney, Charles N., & Ayrs, O. L., 1905b
Safford, James M., 1884
Smith, William G., & Bennett, Hugh H., 1904
Wilder, Henry J., & Geib, W. J., 1904a

Texas
Bennett, Frank, Jr., & Jones, Grove B., 1903
Burgess, James L., & Lyman, W. S., 1907
Caine, Thomas A., & Kocher, A. E., 1904b
Caine, Thomas A., & Lyman, W. S., 1905a
Carter, William T., Jr., & Kocher, A. E., 1905
Carter, William T., Jr., & Kocher, A. E., 1907
Ely, Charles W., & Kocher, A. E., 1908
Hearn, W. Edward, 1904
Hearn, W. Edward, & Burgess, James L., 1904b
Hearn, W. Edward, & Burgess, James L., 1904c
Lapham, J. E., et al, 1903
Lapham, J. E., et al, 1904
Loughridge, R. H., 1884c
Mangum, A. W., & Belden, H. L., 1905
Mangum, A. W., & Carr, M. Earl, 1907
Mangum, A. W., & Lee, Ora, Jr., 1908
Mangum, A. W., & Lyman, W. S., 1908
Martin, J. O., 1902
Mooney, Charles N., et al, 1907

Utah
Gardner, Frank D., & Jensen, Charles A., 1901a
Gardner, Frank D., & Jensen, Charles A., 1901b
Gardner, Frank D., & Stewart, John, 1900
Jensen, Charles A., & Strahorn, A. T., 1905

Means, Thomas H., 1900b
Sanchez, Alfred M., 1904
Vermont
Wilder, Henry J., & Belden, H. L., 1905b
Virginia
Bennett, Frank, Jr., et al, 1908
Bennett, Hugh H., & McLendon, W. E., 1907a
Bennett, Hugh H., & McLendon, W. E., 1907b
Caine, Thomas A., & Bennett, Hugh H., 1905
Carter, William T., Jr., & Lyman, W. S., 1904
Kerr, W. C., 1884b
Lapham, J. E., 1904
Mooney, Charles N., & Bonsteel, F. E., 1903
Mooney, Charles N., & Caine, Thomas A., 1902
Mooney, Charles N.; Martin, F. O., & Caine, Thomas A., 1902
Washington
Carr, E. P., & Mangum, A. W., 1907a
Carr, E. P., & Mangum, A. W., 1907b
Holmes, J. Garnett, 1903a
Jensen, Charles A., 1902
West Virginia
Caine, Thomas A., & Tailby, G. W., Jr., 1908
Griffen, A. M., & Ayrs, Orla L., 1907b
Wisconsin
Bonsteel, Jay A., 1903
Caine, Thomas A., & Lyman, W. S., 1905b
Geib, W. J., & Jones, Grove B., 1907

Jones, Grove B., & Ayrs, Orla L., 1908
Meeker, F. N., & Avon-Burke, R. T., 1907
Smith, William G., 1904
Wyoming
Neill, N. P., et al, 1904
Solution features
Caves
Martel, E. A., 1905
Caves, South Dakota
Gamble, Robert J., 1902
Springs
Arkansas
Haywood, J. K., 1902
Weed, Walter H., 1902
Chile
Gilliss, J. M., 1855
Smith, J. Lawrence, 1855
Geysers
Hayden, F. V., 1872a
Weed, Walter H., 1893
Geysers, Wyoming
Bradley, F. H., 1873
Dana, Edward S., & Grinnell, George B., 1876
Doane, Gustavus C., 1871
Hague, Arnold, 1893a
Hague, Arnold, 1893b
Hayden, F. V., 1883b
Langford, N. P., 1873b
Norris, P. W., 1877
Peale, A. C., 1872
Peale, A. C., 1873
Peale, A. C., 1883
Philippines
McCaskey, H. D., 1907
Thermal springs
Hayden, F. V., 1872a

Geomorphology (Cont'd.)

Thermal springs, Arkansas
Haywood, J. K., 1902
Weed, Walter H., 1902

Thermal springs, California
Loew, Oscar, 1876a

Thermal springs, Colorado
McCauley, C. A. H., 1879
Smart, Charles, 1879

Thermal springs, Japan
Jones, George, 1856a

Thermal springs, Oregon
Dornbach, L. M., 1855

Thermal springs, Wyoming
Bradley, F. H., 1873
Dana, Edward S., & Grinnell, George B., 1876
Doane, Gustavus C., 1871
Hague, Arnold, 1893a
Heizmann, C. L., 1874
Langford, N. P., 1873b
Norris, P. W., 1877
Peale, A. C., 1872
Peale, A. C., 1873
Peale, A. C., 1883

Wyoming
Heizmann, C. L., 1874

Weathering

Massive rocks
Gilbert, G. K., 1905

Geophysics

Observations

Electrical
Matteucci, Carlo, 1871

Georgia

Areal geology
Keith, Arthur, 1902
Smith, Eugene A., 1884a

Economic geology

Gold
Taylor, James W., 1867

Iron
Pumpelly, Raphael, 1886

Mineral resources
Howard, C. W., 1867

Geomorphology

Landform description
Keith, Arthur, 1902
Smith, Eugene A., 1884a

Soils
Smith, Eugene A., 1884a

Soils, Bainbridge
Fippin, Elmer O., & Drake, J. A., 1905a

Soils, Cobb County
Avon-Burke, R. T., & Marean, Herbert W., 1902a

Soils, Covington
Marean, Herbert W., 1902

Soils, Dodge County
Ely, Charles W., & Griffen, A. M., 1905

Soils, Fort Valley
Smith, William G., & Carter, William T., Jr., 1904a

Soils, Spalding County
Lapham, J. E.; Lyman, W. S., & Ely, Charles W., 1907

Soils, Waycross
Carr, M. Earl, & Tharp, W. E., 1908

Index

Germany
  Economic geology
    Education
      Lyman, Benjamin S., 1869
      Ward, Willard P., 1869
    Education, Mining law
      Raymond, Rossiter W., 1869
  Geomorphology
    Plant geography
      Drude, Oscar, 1905
  Glacial geology
    Plant geography
      Drude, Oscar, 1905
  Paleoclimatology
    Quaternary
      Drude, Oscar, 1905
  Surveys
    Geological & topographical surveys
      Comstock, C. B., 1876
    Land & marine surveys
      Wheeler, George M., 1885

Geysers
  Africa, Australia, California, Chile, China-Tibet, Iceland, Indonesia, Japan, Mexico, Nevada, New Zealand, Portugal, Taiwan
    Peale, A. C., 1883
  Iceland, New Zealand
    Weed, Walter H., 1893
  Origin
    Weed, Walter H., 1893
  Wyoming
    Barlow, J. W., 1872
    Bradley, F. H., 1873
    Dana, Edward S., & Grinnell, George B., 1876
    Doane, Gustavus C., 1871
    Hague, Arnold, 1893a
    Hague, Arnold, 1893b
    Hayden, F. V., 1872a
    Hayden, F. V., 1883b
    Langford, N. P., 1873b
    Norris, P. W., 1877
    Peale, A. C., 1872
    Peale, A. C., 1873
    Peale, A. C., 1883
    Weed, Walter H., 1893

Glacial features. See under Geomorphology.

Glacial geology. See also Glacial features under Geomorphology.
  Alaska
    Stikine River glaciers
      Williams, Harry G., 1870
  History
    Favre, Ernest, 1879
  Memoirs
    Agassiz, Louis
      Favre, Ernest, 1879
  Methods
    Agassiz, Louis, 1872a

Glacial lakes. See Glacial features and Lacustrine features under Geomorphology; see Lakes, extinct.

Glaciation. See also Paleoclimatology; Changes in level; Uplifts.
  Causes
    Blytt, A., 1890
    Pilar, George, 1877

Glaciation (Cont'd.)

Deposition and erosion

Quaternary
Wallace, A. R., 1894

Ice movement
Reid, Harry F., 1905b

Isostatic rebound

Maine, Massachusetts, Nova Scotia
Mitchell, Henry, 1880

Ontario, Quebec
Bell, Robert, 1898

Glaciers

Alaska

Muir glacier
Reid, Harry F., 1892

Antarctica
Arctowski, Henryk, 1902

Ice

Flow
Reid, Harry F., 1905b

Surface features, Arctic
Hall, Charles F., 1879

India, Kashmir
Workman, Fanny B., 1905
Workman, W. H., 1905

Lake Superior region
Desor, E., 1851

South America
Holdich, T. H., 1903

Gold. See also Mineral resources.

Alaska
Barnard, E. C., 1899
Brooks, Alfred H., 1899
Brooks, Alfred H.; Richard-son, George B., & Collier, Arthur J., 1901
Eldridge, G. H., & Muldrow, Robert, 1899
Emmons, S. F., 1898
Mendenhall, Walter C., 1899
Mendenhall, Walter C., 1901
Peters, W. J., & Brooks, Alfred H., 1899
Schrader, Frank C., 1899
Schrader, Frank C., & Brooks, Alfred H., 1900
Schrader, Frank C., & Spencer, Arthur C., 1901
Walcott, Charles D., 1897

Andes Mountains, Bolivia, Peru
Gibbon, Lardner, 1854

Brazil
Herndon, Lewis, 1853

California
Ashburner, William, 1867
King, T. Butler, 1850
Mendell, G. H., 1882
Waldeyer, Charles, 1873

Genesis
Bowman, Amos, 1873

Colorado, Minnesota, Montana, New Mexico, Nova Scotia, Saskatchewan, South Dakota, Utah
Taylor, James W., 1867
Taylor, James W., 1868

Crystal structure, Occurrence, Properties
Blake, William P., 1885

Georgia, New Hampshire, North Carolina, South Carolina, Virginia
Taylor, James W., 1867

History, Resources
Blake, William P., 1869
Nevada
Blatchly, A., 1867
Edmunds, J. M., et al, 1863
Nova Scotia
Hamilton, Pierce S., 1868
Oklahoma
Bain, H. Foster, 1904
Panama
Engle, F., 1860
United States
King, Clarence; Emmons, S. F., & Becker, G. F., 1885

Great Basin. See also names of appropriate states.
Areal geology
Blake, William P., 1855a
Emory, William H., 1848
Fremont, John C., 1848
Hall, James, 1852
Johnston, A. R., 1858
Marcou, Jules, 1855a
Marcou, Jules, 1855c
Stansbury, Howard, 1852
Whipple, A. W., 1855
Engineering geology
Railroads
Beckwith, E. G., 1855a
Beckwith, E. G., 1855b
Schiel, James, 1855a
Whipple, A. W., 1855
Geological exploration
Beckwith, E. G., 1855a
Beckwith, E. G., 1855b
Blake, William P., 1855a
Emory, William H., 1848
Fremont, John C., 1848

Gilbert, G. K., 1872
Hall, James, 1852
Hayden, F. V., 1872a
Johnston, A. R., 1858
King, Clarence, 1871
King, Clarence, 1873
King, Clarence, 1874
King, Clarence, 1876
King, Clarence, 1877
Lockwood, Daniel W., 1872
Lyle, D. A., 1872
Marcou, Jules, 1855a
Marcou, Jules, 1855c
Schiel, James, 1855a
Stansbury, Howard, 1852
Wheeler, George M., 1872a
Wheeler, George M., 1872b
Wheeler, George M., 1873
Wheeler, George M., 1874
Wheeler, George M., 1875
Wheeler, George M., 1876
Wheeler, George M., 1877
Wheeler, George M., 1878
Wheeler, George M., 1879a
Whipple, A. W., 1855
Geomorphology
Lacustrine features
Newberry, J. S., 1872
Landform description
Powell, John W., 1878b
Thomas, Cyrus, 1872
U. S. Department of Agriculture, 1871
Glacial geology
Glacial lakes
Newberry, J. S., 1872
Hydrogeology
Hinton, Richard J., 1892
Paleontology
Invertebrata
Hall, James, 1852
Hall, James, 1855
Schiel, James, 1855b

Great Basin (Cont'd.)
Paleobotany
Lesquereux, L., 1872a

Great Lakes region. See also names of individual states.
Glacial geology
Isostatic rebound
Gilbert, G. K., 1899
Geomorphology
Lacustrine features
Gilbert, G. K., 1899

Great Plains. See also names of appropriate states.
Areal geology
Abert, J. W., 1846
Blake, William P., 1855a
Cross, Osborne, 1850
Dodge, G. M., 1869
Emmons, S. F., 1893
Emory, William H., 1848
Engelmann, Henry, 1858
Fremont, John C., 1843
Fremont, John C., 1845
Gregory, J. W., 1892
Hall, James, 1852
Hay, Robert, 1892
Hinton, Richard J., 1890
Lander, Frederick W., 1855
Marcou, Jules, 1855a
Marcou, Jules, 1855c
Marcy, Randolph B., 1853
Shumard, B. F., 1853b
Stansbury, Howard, 1852
Stevens, I. I., 1855a
Stevens, I. I., 1855b
Warren, G. K., 1856
Whipple, A. W., 1855
Economic geology
Coal
Hodge, Joseph T., 1872
Coal, Copper, Gypsum
Marcy, Randolph B., 1853
Mineral resources
Dodge, G. M., 1868
Dodge, G. M., 1869
Mullan, John, 1863
Van Lennep, D., 1868
Engineering geology
Railroads
Beckwith, E. G., 1855a
Beckwith, E. G., 1855b
Dodge, G. M., 1868
Dodge, G. M., 1869
Lander, Frederick W., 1855
Mullan, John, 1863
Schiel, James, 1855a
Stevens, I. I., 1855a
Stevens, I. I., 1855b
Van Lennep, D., 1868
Whipple, A. W., 1855
Roads
Mullan, John, 1863
Geological exploration
Abert, J. W., 1846
Beckwith, E. G., 1855a
Beckwith, E. G., 1855b
Blake, William P., 1855a
Cross, Osborne, 1850
Emory, William H., 1848
Engelmann, Henry, 1858
Fremont, John C., 1843
Fremont, John C., 1845
Hall, James, 1852
Hayden, F. V., 1872b
Lander, Frederick W., 1855
Marcou, Jules, 1855a
Marcou, Jules, 1855c
March, Randolph B., 1853
Mullan, John, 1863
Owen, David D., 1852
Schiel, James, 1855a
Shumard, B. F., 1853b
Stansbury, Howard, 1852
Stevens, I. I., 1855a

Stevens, I. I., 1855b
Warren, G. K., 1856
Warren, G. K., 1858
Whipple, A. W., 1855
Geomorphology
  Lacustrine features
    Newberry, J. S., 1872
  Landform description
    Dodge, G. M., 1869
    Powell, John W., 1878b
    Thomas, Cyrus, 1872
    Van Lennep D., 1868
Glacial geology
  Glacial lakes
    Newberry, J. S., 1872
Hydrogeology
  Gregory, J. W., 1892
  Hay, Robert, 1892
  Hinton, Richard J., 1890
  Hinton, Richard J., 1892
  Nettleton, C. E., 1892
  U. S. Department of Agriculture, 1890
Mineralogy, Petrology
  Hitchock, Edward, 1853
  Shepard, Charles U., 1853
Paleontology
  Invertebrata
    Hall, James, 1852
    Hall, James, 1855
    Meek, F. B., 1872c
    Schiel, James, 1855b
    Shumard, B. F., 1853a
    Shumard, B. F., 1858
  Paleobotany
    Lesquereux, L., 1872a
  Vertebrata
    Leidy, Joseph, 1872b
Hydrodynamics
  Christen, T., 1905

Kansas
  Gregory, J. W., 1892
Nebraska
  Gregory, J. W., 1892
  Hicks, L. E., 1892
New Mexico
  Hill, Robert T., 1892
  Pope, John, 1860
Oklahoma
  Gregory, J. W., 1892
Oregon
  Hinton, Richard J., 1892
Rocky Mountains
  Hinton, Richard J., 1892
  Nettleton, C. E., 1892
South Dakota
  Coffin, Frederick B., 1892
  Culver, Garry E., 1892
Texas
  Hill, Robert T., 1892
  Pope, John, 1858
United States
  Methods
    Hollister, George B., 1905

Greece
  Surveys
    Land and marine surveys
      Wheeler, George M., 1885

Greenland
  Economic geology
    Fluorspar
      Quale, Paul, 1867

Groundwater. See also **Artesian waters and wells**; **Hydrogeology**.

Algeria, Australia, California, Colorado, Great Plains, Kansas, Nebraska, New Mexico, North Dakota, South Dakota, Texas, Utah, Wyoming

U. S. Department of Agriculture, 1890

Algeria, Australia, Great Plains, India, Rocky Mountains

Hinton, Richard J., 1890

Arkansas

Artesian wells

Owen, David D., 1868

California

Hinton, Richard J., 1892
Nutter, Edward H., 1902

Colorado

Gregory, J. W., 1892

Great Basin

Hinton, Richard J., 1892

Great Plains

Gregory, J. W., 1892
Hay, Robert, 1892
Hinton, Richard J., 1892
Nettleton, C. E., 1892

United States, eastern

Methods

Fuller, Myron L., 1905

Washington

Hinton, Richard J., 1892

Guatemala

Earthquakes

Jones, James O., 1903

1857-1858

Canudas, A., 1859

Structural geology

Orogeny, Uplifts

Sapper, Karl, 1905

Volcanology

Jones, James O., 1903
Sapper, Karl, 1905

Gulf of Mexico

Marine geology

Continental shelf

Hilgard, J. E., 1885

Gymnosperms

Arizona, Triassic

Ward, Lester F., 1901

Gypsum

Great Plains, Oklahoma, Texas

Marcy, Randolph B., 1853

Oklahoma

Van Vleet, A. H., 1904

H

Heat flow

Earth cooling

King, Clarence, 1894

History. See also under appropriate subject; see also **Memoirs**.

Coast Survey

Henry, Joseph, 1871

Earth

Pilar, George, 1877

Index

Fifth International Geologic Congress
Congrès Geologique International, 1893
Geochronology
Geikie, Archibald, 1893
Walcott, Charles D., 1894
Geodesy
Merino, Miguel, 1864
1903
Holdich, T. H., 1903
Geological exploration
Baird, Spencer F., 1852
Baird, Spencer F., 1853
Baird, Spencer F., 1855
Geology
Flourens, P., 1863
Gilbert, G. K., 1903
LeConte, Joseph, 1901
Prestwich, Joseph, 1876
Schuchert, Charles, 1905
Stebbins, Rufus P., 1874
1881
Hawes, George W., 1881a
1882
Hunt, T. Sterry, 1884
1883
Hunt, T. Sterry, 1885
1887
McGee, W. J., 1890
1888
McGee, W. J., 1890
United States
Merrill, George P., 1906
Geology and history
Suess, Edward, 1873
Glacial geology
Favre, Ernest, 1879

Gold
Blake, William P., 1869
Memoirs
Agassiz, Louis
Favre, Ernest, 1879
Stebbins, Rufus P., 1874
Bache, A. D.
Henry, Joseph, 1871
Bravais, A.
Elie de Beaumont, L., 1871
Cuvier, G.
Flourens, P., 1869b
de Blainville, D.
Flourens, P., 1866
Dewey, C.
Anderson, Martin B., 1871
Fourier, J.
Arago, D. F. J., 1872
Gray, Asa
Dana, James D., 1890
Farlow, William G., 1890
Hauy, R. J.
Cuvier, Georges, 1861
Huxley, T. H.
Brooks, William K., 1901
Fiske, John, 1901
Gill, Theodore N., 1896
Lartet, Edward
Fischer, P., 1873
Powell, J. W.
Gilbert, G. K., 1903
Saint Hilaire, G.
de Quatrefages, J. L. A., 1863
Flourens, P., 1862
von Buch, L.
Flourens, P., 1863

History (Cont'd.)

von Zittel, K. A.
Schuchert, Charles, 1905

Mineral deposits, genesis
Emmons, S. F., 1905

Mineralogy
Anderson, Martin B., 1871
Cuvier, Georges, 1861
Elie de Beaumont, L., 1871
1881
Hawes, George W., 1881b
1882
Dana, Edward S., 1884
1883
Dana, Edward S., 1885a
1884
Dana, Edward S., 1885b
1885
Dana, Edward S., 1886
1886
Dana, Edward S., 1889
1887-1888
Dana, Edward S., 1890

Mining geology
Browne, J. R., 1867
Raymond, Rossiter W., 1877
Taylor, James W., 1867
Taylor, James W., 1868

Mining law
Raymond, Rossiter W., 1869

North American geology
1886
Darton, Nelson H., 1889

Oceanography
Henry, Joseph, 1871

Paleobotany
Dana, James D., 1890
Farlow, William G., 1890

Paleontology
Arago, D. F. J., 1872
Brooks, William K., 1901
de Quatrefages, J. L. A., 1863
Favre, Ernest, 1879
Fischer, P., 1873
Fiske, John, 1901
Flourens, P., 1862
Flourens, P., 1866
Flourens, P., 1869a
Flourens, P., 1869b
Gill, Theodore N., 1896
Schuchert, Charles, 1905
1887-1888
Williams, Henry S., 1890

Petroleum
Hays, S. S., 1866
Hunt, T. Sterry, 1862
Peckham, S. F., 1884

Petrology
Teall, J. J., Harris, 1903
1887-1888
Merrill, George P., 1890

Platinum
Blake, William P., 1869

Seismology
1884
Rockwood, Charles G., Jr., 1885
1885
Rockwood, Charles G., Jr., 1886
1886
Rockwood, Charles G., Jr., 1889

Silver

Blake, William P., 1869

Smithsonian Institution

Rhees, William J., 1901

Surveys

Baird, Spencer F., 1852
Baird, Spencer F., 1853
Baird, Spencer F., 1855

Minnesota River

Warren, G. K., 1875b

United States

Hayden, F. V., 1878a
House Committee on Public Lands, 1874
Humphreys, A. A., 1878a
Humphreys, A. A., 1878b
McCrary, George W., 1879
Marsh, O. C., 1878
Marsh, O. C., 1879a
Marsh, O. C., 1879b
Powell, John W., 1878a
Powell, John W., 1879
U. S. War Department, 1874
U. S. War Department, 1878
Wheeler, George M., 1879b
Wilson, Joseph S., 1868
Wright, H. G., 1879
Wright, H. G., 1880

U. S. National Museum

Goode, George B., 1901
True, Frederick W., 1898

Volcanology

1884

Rockwood, Charles G., Jr., 1885

1885

Rockwood, Charles G., Jr., 1886

1886

Rockwood, Charles G., Jr., 1889

Holland

Surveys

Land and marine surveys

Wheeler, George M., 1885

Honduras

Areal geology

Davis, C. H., 1903

Economic geology

Mineral resources

Moe, Alfred K., 1904

Engineering geology

Canals, Railroads

Davis, C. H., 1866

Geomorphology

Landform description

Moe, Alfred K., 1904

Structural geology

Orogeny, Uplifts

Sapper, Karl, 1905

Volcanology

Sapper, Karl, 1905

Hungary

Geomorphology

Landform description

de Martonne, E., 1905

Glacial geology

Glacial features

de Martonne, E., 1905

Surveys

Land and marine surveys

Wheeler, George M., 1885

Hydrogeology. See also **Artesian waters and wells; Groundwater.**

Algeria, Australia, India, Rocky Mountains
Hinton, Richard J., 1890

Arkansas
Artesian wells
Owen, David D., 1868

Colorado, Kansas
Gregory, J. W., 1892

Great Plains
Gregory, J. W., 1892
Hay, Robert, 1892
Hinton, Richard J., 1890

Hydrodynamics
Christen, T., 1905

Methods
Fuller, Myron L., 1905
Hollister, George B., 1905

Nebraska
Gregory, J. W., 1892
Hicks, L. E., 1892

New Mexico
Hill, Robert T., 1892
Pope, John, 1860

Nicaragua, Panama
Davis, Arthur P., 1903

Oklahoma
Gregory, J. W., 1892

South Dakota
Coffin, Frederick B., 1892
Culver, Garry E., 1892

Texas
Hill, Robert T., 1892

I

Ice ages. See **Ice ages, ancient;** see **Quaternary.**

Ice ages, ancient
Quaternary
Deposition and erosion
Wallace, A. R., 1894

Iceland
Areal geology
Stefansson, Jon, 1907

Earthquakes
Boehmer, George H., 1886
Stefansson, Jon, 1907

Geomorphology
Geysers
Peale, A. C., 1883
Weed, Walter H., 1893

Volcanology
Boehmer, George H., 1886
Stefansson, Jon, 1907

Idaho
Areal geology
Emmons, S. F., 1893
Hayden, F. V., 1883a

Economic geology
Mineral resources
Ashburner, William, 1869
Browne, J. R., 1867
Browne, J. R., 1868
Mullan, John, 1863
Peale, A. C., 1872
Raymond, Rossiter W., 1869
Raymond, Rossiter W., 1870
Raymond, Rossiter W., 1871

Raymond, Rossiter W., 1872
Raymond, Rossiter W., 1873
Raymond, Rossiter W., 1874
Raymond, Rossiter W., 1875
Raymond, Rossiter W., 1877
Engineering geology
Railroads, Roads
Mullan, John, 1863
Geological exploration
Bradley, F. H., 1873
Hayden, F. V., 1872a
Hayden, F. V., 1873
Hayden, F. V., 1877
Hayden, F. V., 1878b
Hayden, F. V., 1883a
Mullan, John, 1863
Wheeler, George M., 1877
Wheeler, George M., 1878
Wheeler, George M., 1879a
Geomorphology
Landform description
Langford, N. P., 1873a
Soils, Blackfoot
McLendon, W. E., 1904
Soils, Boise
Jensen, Charles A., & Ohlshausen, B. A., 1902
Soils, Lewiston
Mesmer, Louis, 1903
Mineralogy
Peale, A. C., 1872
Paleontology
Invertebrata
Hayden, F. V., 1883a
Meek, F. B., 1873
White, Charles A., 1883

Vertebrata, Tertiary
Cope, E. D., 1884

Igneous rocks
Daubrée, G. A., 1862
Classification, Physical properties
Merrill, George P., 1891
Collections
U. S. National Museum
Merrill, George P., 1891

Illinois
Areal geology, Geological exploration
Locke, John, 1844
Nicollet, I. N., 1843
Owen, David D., 1840
Economic geology
Coal
Jamison, J., 1853
Owen, David D., 1844
Coal, Iron, Lead
Locke, John, 1844
Owen, David D., 1840
Engineering geology
Bridges
Warren, G. K., 1878
Drainage changes
Long, S. H.; Graham, R., & Philips, Joseph, 1819
Roads
Macomb, Alexander; Knight, Jonathan, & Wevers, C. W., 1826
Geomorphology
Fluvial features
Warren, G. K., 1878

Illinois (Cont'd.)

Fluvial features, Illinois River
Long, S. H.; Graham, R., & Philips, Joseph, 1819

Soils
Owen, David D., 1840

Soils, Clay County
Coffey, George N., et al, 1903a

Soils, Clinton County
Bonsteel, Jay A., et al, 1903a

Soils, Johnson County
Coffey, George N., et al, 1904a

Soils, Knox County
Coffey, George N., et al, 1904b

Soils, McLean County
Coffey, George N., et al, 1904c

Soils, O'Fallon
Fippin, Elmer O., & Drake, J. A., 1905b

Soils, St. Clair County
Coffey, George N., et al, 1903b

Soils, Sangamon County
Coffey, George N., et al, 1904d

Soils, Tazewell County
Bonsteel, Jay A., et al, 1903b

Soils, Winnebago County
Coffey, George N., et al, 1904e

Paleontology
Invertebrata
Hayden, F. V., 1883a
Nicollet, I. N., 1843
White, Charles A., 1883

India

Glacial geology
Glacial features, Kashmir
Workman, Fanny B., 1905
Workman, W. H., 1905

Hydrogeology
Hinton, Richard J., 1890

Surveys
Land and marine surveys
Wheeler, George M., 1885

Indiana

Economic geology
Coal
Owen, David D., 1844

Gas, natural
Orton, Edward, 1893

Engineering geology
Roads
Macomb, Alexander; Knight, Jonathan, & Wevers, C. W., 1826

Geomorphology
Soils, Boonville
Mangum, A. W., & Neill, N. P., 1905a

Soils, Greene County
Tharp, W. E., & Mann, Charles J., 1908

Soils, Madison County
Avon-Burke, R. T., & Ruhlen, La Mott, 1904a

Soils, Marshall County
Bennett, Frank, Jr., & Ely, Charles W., 1905
Soils, Newton County
Neill, N. P., & Tharp, W. E., 1907a
Soils, Posey County
Marean, Herbert W., 1903a
Soils, Scott County
Mangum, A. W., & Neill, N. P., 1905b
Soils, Tippecanoe County
Neill, N. P., & Tharp, W. E., 1907b

Paleontology
Invertebrata
Hayden, F. V., 1883a
White, Charles A., 1883
Structural geology
Rock pressure
Orton, Edward, 1893

Indonesia
Geomorphology
Geysers
Peale, A. C., 1883
Paleontology
Man, fossil
Dubois, Eugene, 1899

Insecta
Colorado
Scudder, Samuel H., 1883

Invertebrata. See also phyla and classes.
Prestwich, Joseph, 1876

Arizona
Conrad, T. A., 1857
Hall, James, 1857
Newberry, J. S., 1861
Bibliography, 1884
Marcou, John B., 1885
Bibliography, 1885
Marcou, John B., 1886
Bibliography, 1886
Marcou, John B., 1889
California
Conrad, T. A., 1855a
Conrad, T. A., 1855b
Conrad, T. A., 1855d
Conrad, T. A., 1857
Hall, James, 1845a
Hall, James, 1855
Hall, James, 1857
Newberry, J. S., 1861
Schiel, James, 1855b
Chile
Conrad, T. A., 1855c
Colorado
Hayden, F. V., 1883a
Meek, F. B., 1872c
Scudder, Samuel H., 1885
White, Charles A., 1883
Colorado Plateau
Newberry, J. S., 1861
Great Basin, Great Plains, Rocky Mountains
Hall, James, 1852
Hall, James, 1855
Great Plains
Shumard, B. F., 1853a
Shumard, B. F., 1858
Great Plains, Rocky Mountains
Meek, F. B., 1872c
Schiel, James, 1855b

Invertebrata (Cont'd.)

Idaho
Hayden, F. V., 1883a
Meek, F. B., 1873
White, Charles A., 1883

Illinois
Hayden, F. V., 1883a
Nicollet, I. N., 1843
White, Charles A., 1883

Indiana
Hayden, F. V., 1883a
White, Charles A., 1883

Iowa
Hayden, F. V., 1883a
Nicollet, I. N., 1843
Owen, David D., 1848b
White, Charles A., 1883

Kansas
Shumard, B. F., 1858

Lake Superior region
Locke, John, 1849
Owen, David D., 1848b

Lake Superior region, Michigan
Hall, James, 1851a
Locke, John, 1848

Louisiana, Recent
Hilgard, Eugene W., & Hopkins, F. V., 1878

Mexico
Baja California
Newberry, J. S., 1861

Baja California, Chihuahua, Coahuila, Nuevo Leon, Sonora, Tamaulipas
Conrad, T. A., 1857
Hall, James, 1857

Minnesota
Hall, James, 1851a
Owen, David D., 1848b

Mississippi Valley
Nicollet, I. N., 1843

Missouri
Hayden, F. V., 1883a
Nicollet, I. N., 1843
White, Charles A., 1883

Mollusca
Carpenter, Philip P., 1861

Montana
Meek, F. B., 1873
Whitfield, R. P., 1876

Nebraska
Meek, F. B., 1871
Meek, F. B., 1872c
Shumard, B. F., 1858

Nevada
Newberry, J. S., 1861

New Mexico
Abert, J. W., 1848
Bailey, J. W., 1848
Conrad, T. A., 1857
Cope, E. D., 1875
Hall, James, 1857

North America, 1884
Marcou, John B., 1885

North America, 1885
Marcou, John B., 1886

North America, 1886
Marcou, John B., 1889

Oklahoma
Shumard, B. F., 1853a

Oregon
Conrad, T. A., 1855b
Hall, James, 1845a

Index

South Dakota
Grinnell, George B., 1874
Grinnell, George B., 1875
Meek, F. B., 1872c
Shumard, B. F., 1858

Texas
Conrad, T. A., 1857
Hall, James, 1857
Ruffner, E. H., 1877b
Shumard, B. F., 1853a

United States

Southwest
Conrad, T. A., 1857
Hall, James, 1857

Western
Hall, James, 1855

Utah
Hall, James, 1852

Utah, Wyoming
Hayden, F. V., 1883a
Meek, F. B., 1872b
Meek, F. B., 1872c
Meek, F. B., 1873
White, Charles A., 1883

Wisconsin
Hall, James, 1851a
Nicollet, I. N., 1843
Owen, David D., 1848b

Wyoming
Shumard, B. F., 1858

Iowa

Areal geology
Allen, J., 1846
Locke, John, 1844
Nicollet, I. N., 1843
Owen, David D., 1840
Owen, David D., 1848a
Owen, David D., 1848b

Economic geology

Coal, Iron
Owen, David D., 1844

Copper
Owen, David D., 1848a

Iron, Lead, Zinc
Locke, John, 1844
Owen, David D., 1840

Engineering geology

Bridges
Warren, G. K., 1878

Geological exploration
Allen, J., 1846
Locke, John, 1844
Nicollet, I. N., 1843
Owen, David D., 1840

Geomorphology

Fluvial features
Warren, G. K., 1878

Glacial features
Owen, David D., 1848b

Soils
Owen, David D., 1840

Soils, Cerro Gordo County
Marean, Herbert W., &
  Jones, Grove B., 1904a

Soils, Dubuque
Fippin, Elmer O., 1903a

Soils, Story County
Marean, Herbert W., &
  Jones, Grove B., 1904b

Soils, Tama County
Ely, Charles W.; Coffey,
  George N., & Griffen,
  A. M., 1905

Iowa (Cont'd.)

Paleontology
Invertebrata
Hayden, F. V., 1883a
Nicollet, I. N., 1843
Owen, David D., 1848b
White, Charles A., 1883

Ireland

Geomorphology
Plant distribution
Anderson, Richard J., 1905
Paleontology
Paleobotany, Quaternary
Anderson, Richard J., 1905

Iron

Alabama
Alabama, University of, 1902
Alabama, Connecticut, Georgia, Kentucky, Maine, Maryland, Massachusetts, Michigan, Minnesota, Missouri, New Hampshire, New Jersey, New York, North Carolina, Ohio, Pennsylvania, Tennessee, Vermont, Virginia, Wisconsin
Pumpelly, Raphael, 1886
Arkansas
Linn, L. F., & Sevier, A. H., 1836
Geographic distribution, Geologic distribution
Pumpelly, Raphael, 1886
Illinois, Iowa, Wisconsin
Locke, John, 1844
Owen, David D., 1840

Lake Superior region
Foster, J. W., & Whitney, J. D., 1851
Jackson, Charles T., 1849
Michigan, Minnesota, Wisconsin
Foster, J. W., & Whitney, J. D., 1851
North Carolina
Laidley, T. T. S., 1856
Wilkes, Charles; Hunt, H., & Martin, D. B., 1859
Reserves
Leith, Charles K., 1907
United States, western
Pumpelly, Raphael, 1886

Isostasy. See also Changes in level.

Gilbert, G. K., 1893
Geomorphologic effects
Hudson Bay, Ontario, Quebec
Bell, Robert, 1898
Great Lakes region
Gilbert, G. K., 1899

Israel

Areal geology, Geological exploration
Dead Sea
Lynch, W. F., 1848
Geomorphology
Landform description
Lynch, W. F., 1848

Italy

Surveys

Geological and topographical surveys
Comstock, C. B., 1876

Land and marine surveys
Wheeler, George M., 1885

Volcanology

Pompeii
Heilprin, Angelo, 1905a

Vesuvius
Lacroix, A., 1907

J

Japan

Areal geology
Jones, George, 1856b
Perry, M. C., 1856

Earthquakes

1854
Bache, A. D., 1856a

Geomorphology

Geysers, thermal springs
Jones, George, 1856a
Peale, A. C., 1883

Marine geology

Tsunamis
Bache, A. D., 1856a

Mineralogy

Mineral waters
Jones, George, 1856a

Surveys

Land and marine surveys
Wheeler, George M., 1885

K

Kansas

Areal geology
Emmons, S. F., 1893
Engelmann, Henry, 1858
Gregory, J. W., 1892
Hayden, F. V., 1858

Economic geology

Mineral resources
Elliott, R. S., 1872

Geological exploration
Engelmann, Henry, 1858
Hayden, F. V., 1858
Warren, G. K., 1858
Wheeler, George M., 1878
Wheeler, George M., 1879a

Geomorphology

Soils, Allen County
Drake, J. A., & Tharp, W. E., 1905

Soils, Brown County
Burgess, James L.; Tharp, W. E., & Lymann, W. S., 1907

Soils, Garden City
Burgess, James L., & Coffey, George N., 1905

Soils, Parsons
Drake, J. A., 1904

Soils, Riley County
Carter, William T., Jr., & Smith, Howard C., 1908

Soils, Russell
Mangum, A. W., & Drake, J. A., 1904

Soils, Wichita
Lapham, J. E., & Olshausen. B. A., 1903

Kansas (Cont'd.)

Hydrogeology
Gregory, J. W., 1892
U. S. Department of Agriculture, 1890

Paleontology
Hayden, F. V., 1858

Invertebrata
Shumard, B. F., 1858

Man, fossil
Holmes, William H., 1903b

Paleobotany
Lesquereux, L., 1872b

Vertebrata
Cope, E. D., 1872b
Cope, E. D., 1872c
Cope, E. D., 1884

Keewatin

Geomorphology
Landform description
Wilson, A. W. G., 1905

Kentucky

Areal geology
Brown, C. Newton, 1900

Economic geology
Coal
Brown, C. Newton, 1900
Jamison, J., 1853
Owen, David D., 1844
Shaler, N. S., 1875

Iron
Pumpelly, Raphael, 1886

Geomorphology
Soils, McCracken County
Rice, Thomas D., 1907

Soils, Madison County
Griffen, A. M., & Ayrs, Orla L., 1907a

Soils, Mason County
Avon-Burke, R. T., 1904a

Soils, Scott County
Avon-Burke, R. T., 1904b

Soils, Union County
Marean, Herbert W., 1903b

Soils, Warren County
Rice, Thomas D., & Geib, W. J., 1905d

L

Labrador

Geomorphology
Landform description
Lieber, Oscar M., 1861
Wilson, A. W. G., 1905

Lake Superior region. See also names of individual states.

Areal geology
Allen, J., 1834
Barnes, George O., 1849a
Barnes, George O., 1849b
Burt, William A., 1846
Burt, William A., & Hubbard, Bela, 1849
Channing, William F., 1848a
Channing, William F., 1848b
Channing, William F., 1848c
Congrès Geologique International, 1893
Dickenson, George J., 1849
Featherstonhaugh, George W., 1836
Foster, J. W., 1848

Foster, J. W., 1849a
Foster, J. W., 1849b
Foster, J. W., & Whitney,
  J. D., 1849
Foster, J. W., & Whitney,
  J. D., 1850a
Foster, J. W., & Whitney,
  J. D., 1850b
Foster, J. W., & Whitney,
  J. D., 1851
Gibbs, William P., 1849
Hubbard, Bela, 1846
Jackson, Charles T., 1848a
Jackson, Charles T., 1848b
Jackson, Chrles T., 1849
Locke, John, 1848
McIntyre, James, 1849
Norwood, J. G., 1848
Owen, David D., 1848a
Owen, David D., 1848b
Pope, John, 1850
Van Hise, C. R., 1893
Whitney, J. D., 1848a
Whitney, J. D., 1848b
Whitney, J. D., 1849

Economic geology

Copper

Allen, J., 1834
Bell, William H., 1844
Burt, William A., 1846
Cass, Lewis, 1825
Channing, William F.,
  1848a
Channing, William F.,
  1848b
Channing, William F.,
  1848c
Foster, J. W., 1848
Foster, J. W., & Whitney,
  J. D., 1850b
Gray, A. B., 1846
Houghton, Douglas, 1832
Hubbard, Bela, 1846
Jackson, Charles T., 1848a
Jackson, Charles T., 1848b
Jackson, Charles T., 1849

Locke, John, 1848
Owen, David D., 1848a
Schoolcraft, Henry R., 1822
Stockton, John, et al, 1845
Whitney, J. D., 1848a
Whitney, J. D., 1848b

Iron

Foster, J. W., & Whitney,
  J. D., 1851
Jackson, Charles T., 1849

Mineral resources

Featherstonhaugh, George
  W., 1836
Foster, J. W., & Whitney,
  J. D., 1850a

Silver

Jackson, Charles T., 1849

Geological exploration

Allen, J., 1834
Featherstonhaugh, George
  W., 1836
Foster, J. W., & Whitney,
  J. D., 1850b
Jackson, Charles T., 1849
Pope, John, 1850

Geomorphology

Glacial features

Owen, David D., 1848b

Landform description

Foster, J. W., 1849b

Terraces

Desor, E., 1851

Glacial geology

Drift, Glacial lakes

Desor, E., 1851

Mineralogy, Petrology

Burt, William A., & Hubbard, Bela, 1849
Locke, John, 1849

Lake Superior region (Cont'd.)

Paleontology

Invertebrata
Hall, James, 1851a
Locke, John, 1848
Locke, John, 1849
Owen, David D., 1848b

Stratigraphy
Hall, James, 1851b
Hall, James, 1851c

Lakes. See also **Lacustrine features** under **Geomorphology.**

Description

Florida
MacGonigle, John N., 1905

Shoreline changes

Lake Michigan
Bixby, W. H.; Beach, Lansing H., & Gaillard, D. D., 1906

Lakes, extinct. See also **Lacustrine features** under **Geomorphology.**

Great Basin, Great Plains, Rocky Mountains
Newberry, J. S., 1872

New York, Ontario, Quebec St. Lawrence Valley
Coleman, A. P., 1905

Landforms. See **Landform description** and **Landform evolution** under **Geomorphology.**

Lead

England, Wisconsin
Bell, William H., 1844

Illinois, Iowa, Wisconsin
Locke, John, 1844
Owen, David D., 1840

New Mexico
Taylor, James W., 1867
Taylor, James W., 1868

Life, origin. See under **Paleontology.**

Limestone. See also under **Sedimentary rocks.**

Alabama
Eckel, Edwin C., 1903
Eckel, Edwin C., & Crider, A. F., 1905
Smith, Eugene A., 1903

California
Williamson, R. S., 1850

Cement manufacture
Eckel, Edwin C., 1903

Mississippi
Eckel, Edwin C., & Crider, A. F., 1905

Mississippi Valley
U. S. Department of Agriculture, 1870

Oregon
Frazer, John F., 1850

United States, eastern
U. S. Department of Agriculture, 1869c

United States, southern
U. S. Department of Agriculture, 1869a

Virginia
U. S. Department of Agriculture, 1869b

Index 211

Lineation
Hobbs, William H., 1905

Loess. See under **Sediments.**

Louisiana
Areal geology, Geological exploration
Blake, William P., 1855c
Featherstonhaugh, George W., 1835
Hilgard, Eugene W., 1884c
Marcou, Jules, 1855b
Pope, John, 1855

Economic geology
Iron
Johnson, Lawrence C., 1888

Engineering geology
Canals
U. S. Army Corps of Engineers, 1874

Floods
Ellet, Charles, Jr., 1852

Lake Borgne & Mississippi River
Forshey, C. G., 1875

Railroads
Pope, John, 1855

Geomorphology
Deltas
Delafield, Richard, 1824
Ellet, Charles, Jr., 1851
Ellet, Charles, Jr., 1852
Graham, George, 1829
Forshey, C. G., 1875
Hilgard, Eugene W., & Hopkins, F. V., 1878
Long, S. H., & Humphreys, A. A., 1851
U. S. Army Corps of Engineers, 1874

Fluvial features
Graham, George, 1829
Hilgard, Eugene W., & Hopkins, F. V., 1878
Little, George, 1882
U. S. Army Corps of Engineers, 1874

Fluvial features, Lacustrine features
Forshey, C. G., 1875

Landform description
Hilgard, Eugene W., 1884c
Little, George, 1882

Log dams, Red River
Fuller, Charles A., 1855
Linnard, T. B., 1844
Long, S. H., 1841
Paxton, Joseph, 1829
Shreve, Henry M., 1833
Shreve, Henry M., & Gratiot, C., 1834

Loess
Little, George, 1882

Mississippi River
Ellet, Charles, Jr., 1851
Ellet, Charles, Jr., 1852
Graham, George, 1829
Long S. G., & Humphreys, A. A., 1851

Mudlumps
Delafield, Richard, 1829

Soils
Hilgard, Eugene W., 1884c

Soils, Acadia Parish
Rice, Thomas D., & Griswold, Lewis, 1904a

Soils, Caddo Parish
Burgess, James L., et al, 1908

Louisiana (Cont'd.)

Soils, De Soto Parish
Jones, Grove B., & Ruhlen, La Mott, 1905

Soils, East Baton Rouge Parish
Ely, Charles W.; Marean, Herbert W., & Neill, N. P., 1907

Soils, Lake Charles
Heileman, W. H., & Mesmer, Louis, 1902

Soils, New Orleans
Rice, Thomas D., & Griswold, Lewis, 1904b

Soils, Ouachita Parish
Rice, Thomas D., 1904

Soils, Tangipahoa Parish
Griffen, A. M., & Caine, Thomas A., 1907

Marine geology
Shore features
U. S. Army Corps of Engineers, 1874

Shoreline sediments
Hilgard, Eugene W., & Hopkins, F. V., 1878

Paleontology
Diatoms, Foraminifera.
Hilgard, Eugene W., & Hopkins, F. V., 1878

Stratigraphy
Quaternary
Little, George, 1882

M

Mackenzie
Geomorphology
Landform description
Wilson, A. W. G., 1905

Magnetic field, earth
Description
Creak, Ettrick W., 1904
Variations
Volcanism
Bauer, L. A., 1905

Maine
Economic geology
Iron
Pumpelly, Raphael, 1886
Geomorphology
Shore features
Mitchell, Henry, 1880
Glacial geology
Isostatic rebound
Mitchell, Henry, 1880
Marine geology
Shore features
Mitchell, Henry, 1880

Malacostraca
Carboniferous
Packard, A. S., 1886d

Mammalia
Chile
Wyman, Jeffries, 1855
Europe, Quaternary
Rütimeyer, L., 1862

Great Plains, Tertiary
   Leidy, Joseph, 1872b
Nebraska
   Leidy, Joseph, 1852
New Mexico, Tertiary
   Cope, E. D., 1874
   Cope, E. D., 1875
Oklahoma, Quaternary
   Holmes, William H., 1903a
Quaternary
   Lucas, Frederick A., 1901b
   Lydekker, R., 1901
Restoration
   Elephas primigenius
   Pfizenmayer, E., 1907
Rocky Mountains, Tertiary
   Leidy, Joseph, 1872b
U. S. S. R., Quaternary
   Herz, O. F., 1904
Utah, Tertiary
   Cope, E .D., 1872d
Wyoming, Tertiary
   Cope, E. D., 1873
   Leidy, Joseph, 1872a

Man, fossil
   California, Quaternary
   Holmes, William H., 1901
   Europe
   Obermaier, Hugues, 1906
   Seuss, Edward, 1873
   Evolution
      Foot
      Anthony, Raoul, 1904
   France
   Gaudry, Albert, 1903

Indonesia
   Dubois, Eugene, 1899
Kansas
   Holmes, William H., 1903b
Origin
   Haeckel, Ernst, 1899
Pithecanthropus erectus
   Dubois, Eugene, 1899

Manchuria
   Geological exploration
   Collins, P. McD., 1858

Manitoba
   Geomorphology
   Landform description
   Wilson, A. W. G., 1905

Mantle
   Structure
   Peirce, Benjamin, 1881

Marine geology. See also **Oceanography**; **Continental shelf**.
   Abyssal fauna
   Origin
   Ortmann, A. E., 1905
   Bottom features
   Murray, John, 1901
   Thoulet, J., 1905
   Atlantic Ocean
   Packard, A. S., 1905
   Caribbean Sea
   Agassiz, Alexander, 1881
   Pacific Ocean
   Clover, Richardson, 1892

Marine geology (Cont'd.)

United States
Hilgard, J. E., 1885

Foraminifera

Atlantic Ocean
Pourtales, L. F., 1859

Instruments
Craven, T. A., 1855
Mitchell, Henry, 1861
Sands, B. F., 1856
Trowbridge, W. P., 1860

Ocean basins

Distribution
Gregory, J. W., 1899

Origin
Kelvin, William T., 1898
Murray, John, 1901

Reefs

Description, Origin
Sollas, W. J., 1899

Florida
Agassiz, Louis, 1852
Hunt, E. B., 1864
Pourtales, L. F., 1871

Sea level changes
Blytt, A., 1890

Sediments

Abyssal
Daubrée, A., 1894
Murray, John, 1901
Murray, John, 1905

Atlantic Ocean
Agassiz, Louis, 1872b
Bailey, J. W., 1856a
Bailey, J. W., 1856b
Pourtales, L. F., 1854
Pourtales, L. F., 1859
Pourtales, L. F., 1872

Caribbean Sea
Agassiz, Alexander, 1881

Chesapeake Bay
Hughes, George W., 1837

Connecticut, Delaware, Massachusetts, New Jersey, New York, Rhode Island.
Lindenkohl, A., 1885

Florida
Agassiz, Louis, 1852
Hunt, E. B., 1864

Louisiana, Mississippi delta.
Hilgard, Eugene W., & Hopkins, F. V., 1878

Marginal
Daubrée, A., 1894

Pacific Ocean
Clover, Richardson, 1892

Shore features. See also **Shorelines; Coastal features** under **Geomorphology.**

Delaware
Mitchell, Henry, 1889a

Florida
Gibbs, Oliver W., 1856

Louisiana
U. S. Army Corps of Engineers, 1874

Maine, Nova Scotia
Mitchell, Henry, 1880

Massachusetts
Boutelle, Charles O., 1887
Gulliver, F. P., 1905
Marinden, Henry L., 1890a
Marinden, Henry L., 1890b
Marinden, Henry L., 1892a
Marinden, Henry L., 1892b
Marinden, Henry L., 1894
Marinden, Henry L., 1896

Marinden, Henry L., 1897
Mitchell, Henry, 1874
Mitchell, Henry, 1875
Mitchell, Henry, 1880
Mitchell, Henry, 1887a
Mitchell, Henry, 1889b
Van Ingen, H. S., 1874
Whiting, Henry L., 1875
Whiting, Henry L., 1887
Whiting, Henry L., 1890
Methods
Agassiz, Louis, 1868
New Jersey
Mitchell, Henry, 1889a
New York
Bache, A. D., 1856b
Nicaragua
Mitchell, Henry, 1877
North Carolina
U. S. Army Corps of Engineers, 1905
Oregon, Washington
Langfitt, W. C., 1903
Panama
Mitchell, Henry, 1877
Tidal features
Hilgard, J. E., 1875
Tsunamis
California, Japan
Bache, A. D., 1856a
California, 1868
Hilgard, J. E., 1872

Maryland
Areal geology
McGee, W. J., et al, 1893
Economic geology
Iron
Pumpelly, Raphael, 1886

Engineering geology
Dredging
Hughes, George W., 1837
Geomorphology
Soils, Calvert County
Bonsteel, Jay A., & Avon-Burke, R. T., 1901
Soils, Cecil County
Dorsey, Clarence W., & Bonsteel, Jay A., 1901
Soils, Hartford County
Smith, William G., & Martin, J. O., 1902
Soils, Kent County
Bonsteel, Jay A., 1901a
Soils, Prince George County
Bonsteel, Jay A., et al, 1902a
Soils, St. Mary County
Bonsteel, Jay A., 1901b
Soils, Worcester County
Bonsteel, F. E., & Carter, William T., Jr., 1904
Marine geology
Sediments, Chesapeake Bay, Susquehanna River
Hughes, George W., 1837

Massachusetts
Economic geology
Iron
Pumpelly, Raphael, 1886
Engineering geology
Shorelines
Van Ingen, H. S., 1874
Geomorphology
Fluvial features
Powell, Charles F., 1905

Massachusetts (Cont'd.)

Shore features
Boutelle, Charles O., 1887
Marinden, Henry L., 1890a
Marinden, Henry L., 1890b
Marinden, Henry L., 1892a
Marinden, Henry L., 1892b
Marinden, Henry L., 1894
Marinden, Henry L., 1896
Marinden, Henry L., 1897
Mitchell, Henry, 1874
Mitchell, Henry, 1875
Mitchell, Henry, 1880
Mitchell, Henry, 1887a
Mitchell, Henry, 1889b
Van Ingen, H. S., 1874
Whiting, Henry L., 1875
Whiting, Henry L., 1887
Whiting, Henry L., 1890

Soils, Connecticut Valley
Fippin, Elmer O., 1904

Glacial geology

Isostatic rebound
Mitchell, Henry, 1880

Marine geology

Continental shelf
Lindenkohl, A., 1885

Sediments, Shore features
Mitchell, Henry, 1889b

Shore features
Boutelle, Charles O., 1887
Marinden, Henry L., 1890a
Marinden, Henry L., 1890b
Marinden, Henry L., 1892a
Marinden, Henry L., 1892b
Marinden, Henry L., 1894
Marinden, Henry L., 1896
Marinden, Henry L., 1897
Mitchell, Henry, 1874
Mitchell, Henry, 1875
Mitchell, Henry, 1880
Mitchell, Henry, 1887a

Van Ingen, H. S., 1874
Whiting, Henry L., 1875
Whiting, Henry L., 1887
Whiting, Henry L., 1890

Meanders. See under **Rivers**; see under **Fluvial features** under **Geomorphology.**

Memoirs

Agassiz, Louis
Favre, Ernest, 1879
Stebbins, Rufus P., 1874

Bache, Alexander D.
Henry, Joseph, 1871

Bravais, Auguste
Elie de Beaumont, L., 1871

Cuvier, Georges
Flourens, P., 1869b

de Blainville, D.
Flourens, P., 1866

Dewey, C.
Anderson, Martin B., 1871

Fourier, Joseph
Arago, D. F. J., 1872

Gray, Asa
Dana, James D., 1890
Farlow, William G., 1890

Hauy, Rene-Just
Cuvier, Georges, 1861

Huxley, T. H.
Brooks, William K., 1901
Fiske, John, 1901
Gill, Theodore N., 1896

Lartet, Edward
Fischer, P., 1873

Powell, J. W.
  Gilbert, G. K., 1903
Saint Hilaire, G.
  de Quatrefages, J. L. A., 1863
  Flourens, P., 1862
von Buch, Leopold
  Flourens, P., 1863
von Zittel, K. A., 1905
  Schuchert, Charles, 1905

Mercury
  Andes Mountains, Bolivia, Peru
    Gibbon, Lardner, 1854

Metals
  Origin
    Kemp, James F.,1907
  U. S. S. R., Siberia
    Palmer, Aaron H., 1848

Metamorphic rocks
  Daubrée, G. A., 1862
  Classification, Physical properties
    Merrill, George P., 1891
  Collections
    U. S. National Museum
      Merrill, George P., 1891

Metamorphism
  Daubrée, G. A., 1862

Meteorites
  Smith, J. Lawrence, 1856

Collections
  U. S. National Museum
    Clarke, F. W., 1889
    Tassin, Wirt, 1902b
  Yale University
    Brush, G. J., 1869
Composition, Origin
  Daubrée, G. A., 1869
Description
  Brenndecke, F., 1871
  Chile
    Philippi, R. A., 1855
  Mexico, Chihuahua
    Pierson, William M., 1874

Mexico
  Areal geology
    Davis, C. H., 1866
  Baja California
    Chandler, M. T. W., 1857
    Conrad, T. A., 1857
    Derby, George H., 1852a
    Emory, William H., 1857
    Hall, James, 1857
    Ives, Joseph C., 1861
    Michler, N., 1857b
    Newberry, J. S., 1861
    Parry, C. C., & Schott, Arthur, 1857
  Baja California, Baja California Sur
    Merrill, George P., 1897
  Chihuahua, Coahuila, Nuevo Leon, Tamaulipas
    Conrad, T. A., 1857
    Emory, William H., 1857
    Hall, James, 1857
    Michler, N., 1857a
    Parry, C. C., & Schott, Arthur, 1857

Mexico (Cont'd.)
Wislezenus, A., 1848
Oaxaca, Vera Cruz
Spear, John C., 1872
Sonora
Conrad, T. A., 1857
Emory, William H., 1857
Hall, James, 1857
Michler, N., 1857b
Parry, C. C., & Schott, Arthur, 1857
Earthquakes
1866
Sartorius, Charles, 1867
Economic geology
Mineral resources
Bureau of the American Republics, 1904b
Mineral resources, Sonora
Fergusson, D., 1863
Engineering geology
Canals
Davis, C. H., 1866
Canals, Oaxaca, Vera Cruz
Fuertes, E. A., 1872
Shufeldt, Robert W., 1872
Railroads
Davis, C. H., 1866
Roads, Sonora
Fergusson, D., 1863
Geological exploration
Baja California
Chandler, M. T. W., 1857
Derby, George H., 1852a
Emory, William H., 1857
Ives, Joseph C., 1858
Ives, Joseph C., 1861
Michler, N., 1857b
Newberry, J. S., 1861

Parry, C. C., & Schott, Arthur, 1857
Chihuahua, Coahuila
Emory, William H., 1857
Michler, N., 1857a
Parry, C. C., & Schott, Arthur, 1857
Wislezenus, A., 1848
Nuevo Leon, Tamaulipas
Emory, William H., 1857
Parry, C. C., & Schott, Arthur, 1857
Wislezenus, A., 1848
Oaxaca, Vera Cruz
Shufeldt, Robert W., 1872
Spear, John C., 1872
Sonora
Emory, William H., 1857
Fergusson, D., 1863
Michler, N., 1857b
Parry, C. C., & Schott, Arthur, 1857
Geomorphology
Geysers
Peale, A. C., 1883
Landform description
Bureau of the American Republics, 1904b
Hill, Robert T., 1905a
Landform description, Baja California, Baja California Sur
Merrill, George P., 1897
Mineralogy
Meteorites, Chihuahua
Pierson, William M., 1874
Paleontology
Baja California
Newberry, J. S., 1861

Baja California, Chihuahua, Coahuila, Sonora, Tamaulipas
Conrad, T. A., 1857
Hall, James, 1857

Nuevo Leon
Hall, James, 1857

Structural geology
Orogeny and uplifts
Sapper, Karl, 1905

Volcanology
Sapper, Karl, 1905

Colima
Sartorius, Charles, 1871

Michigan
Areal geology
Burt, William A., & Hubbard, Bela, 1849
Channing, William F., 1848a
Channing, William F., 1848b
Channing, William F., 1848c
Foster, J. W., 1849b
Foster, J. W., & Whitney J. D., 1849
Foster, J. W., & Whitney, J. D., 1850b
Foster, J. W., & Whitney, J. D., 1851
Jackson, Charles T., 1848a
Locke, John, 1848
Whitney, J. D., 1848a

Economic geology
Copper
Channing, William F., 1848a
Channing, William F., 1848b
Channing, William F., 1848c
Foster, J. W., & Whitney, J. D., 1850b
Jackson, Charles T., 1848a
Locke, John, 1848
Whitney, J. D., 1848a

Iron
Foster, J. W., & Whitney, J. D., 1851
Pumpelly, Raphael, 1886

Geological exploration
Foster, J. W., & Whitney, J. D., 1850b

Geomorphology
Lacustrine features
Bixby, W. H.; Beach, Lansing H., & Gaillard, D. D., 1906

Landform description
Foster, J. W., 1849b

Soils, Allegan County
Fippin, Elmer O., & Rice, Thomas D., 1902

Soils, Alma
Hearn, W. Edward, & Griffen, A. M., 1905

Soils, Cass County
Geib, W. J., 1908

Soil, Munising
Rice, Thomas D., & Geib, W. J., 1905c

Soils, Owosso
Mangum, A. W., & Mann, Charles J., 1905

Soils, Oxford
Jones, Grove B., & Carr, M. Earl, 1907

Michigan (Cont'd.)

Soils, Pontiac
Wilder, Henry J., & Geib, W. J., 1904b

Soils, Saginaw
McLendon, W. E., & Carr, M. Earl, 1905

Terraces
Desor, E., 1851

Glacial geology

Drift, Glacial lakes
Desor, E., 1851

Mineralogy, Petrology
Burt, William A., & Hubbard, Bela, 1849

Paleontology

Invertebrata
Hall, James, 1851a
Locke, John, 1848

Stratigraphy
Hall, James, 1851b

Mineral deposits

Classification
Raymond, Rossiter W., 1870

Mineral deposits, genesis

Emmons, S. F., 1905

California, Gold
Bowman, Amos, 1873

California, Sierra Nevada
Bowman, Amos, 1875
Raymond, Rossiter W., 1875

Coal
U. S. Department of Commerce and Labor, 1905

History
Emmons, S. F., 1905

Lake Superior region
Jackson, Charles T., 1849

Metals
Kemp, James F., 1907

Petroleum
U. S. Department of Commerce and Labor, 1905

Mineral economics

Copper

Lake Superior region
Houghton, Douglas, 1832
Schoolcraft, Henry R., 1822

Mineral resources

United States, western
Browne, J. R., 1867
Browne, J. R., 1868

Mineral exploration. See also Geological exploration.

Lake Superior region, Michigan
Channing, William F., 1848a
Channing, William F., 1848b
Channing, William F., 1848c

Lake Superior region, Copper
Allen, J., 1834
Houghton, Douglas, 1832
Schoolcraft, Henry R., 1822

United States
Featherstonhaugh, George W., 1833

Mineral resources. See also **Resources** under **Economic geology**; see also names of individual commodities.

Alabama
Walcott, Charles D., 1901
Walcott, Charles D., 1905

Alaska
Abercrombie, W. R., 1900
Blake, Theodore A., 1869
Browne, J. R., 1867
Browne, J. R., 1868
Bryant, Charles, 1870
Bulkley, Charles S., 1868
Davidson, George, 1868
Eldridge, G. H., 1898b
Eldridge, G. H., 1899
McIntyre, H. H., 1870
Mendenhall, Walter C., 1898a
Mendenhall, Walter C., 1898b
Mendenhall, Walter C., 1898c
Morris, William G., 1879
Raymond, Charles W., 1871
Raymond, Rossiter W., 1874
Schrader, Frank C., 1898
Senate Committee on Military Affairs, 1900
Spurr, J. E., & Post, W. S., 1899
Williams, Harry G., 1870

Algeria
Gautier, E. F., 1905

Argentina
Bureau of the American Republics, 1903

Arizona
Browne, J. R., 1867
Browne, J. R., 1868

Fergusson, D., 1863
Raymond, Rossiter W., 1869
Raymond, Rossiter W., 1870
Raymond, Rossiter W., 1871
Raymond, Rossiter W., 1872
Raymond, Rossiter W., 1873
Raymond, Rossiter W., 1874
Raymond, Rossiter W., 1875
Raymond, Rossiter W., 1877

Bolivia
Bureau of the American Republics, 1904a

Brazil
Bureau of the American Republics, 1902

British Columbia
Blake, Theodore A., 1869

California
Brown, Edwin C., 1905
Browne, J. R., 1867
Browne, J. R., 1868
Preston, E. B., 1905
Raymond, Rossiter W., 1869
Raymond, Rossiter W., 1870
Raymond, Rossiter W., 1871
Raymond, Rossiter W., 1872
Raymond, Rossiter W., 1873
Raymond, Rossiter W., 1874
Raymond, Rossiter W., 1875

Mineral resources (Cont'd.)

Raymond, Rossiter W., 1877
Tyson, P. T., 1850

Caribbean region

Isle of Pines
U. S. War Department, 1906a
U. S. War Department, 1906b

Colorado
Elliott, R. S., 1872
Hawn, F., 1874d
McCauley, C. A. H., 1878
Raymond, Rossiter W., 1870
Raymond, Rossiter W., 1871
Raymond, Rossiter W., 1872
Raymond, Rossiter W., 1873
Raymond, Rossiter W., 1874
Raymond, Rossiter W., 1875
Raymond, Rossiter W., 1877

Dominican Republic
Blake, William P., 1871c
McClellan, George B., 1871
Marvine, A. R., 1871

Georgia
Howard, C. W., 1867

Great Plains
Dodge, G. M., 1868
Dodge, G. M., 1869
Mullan, John, 1863
Van Lennep, D., 1868

Honduras
Moe, Alfred K., 1904

Idaho
Ashburner, William, 1869
Browne, J. R., 1867
Browne, J. R., 1868
Mullan, John, 1863
Peale, A. C., 1872
Raymond, Rossiter W., 1869
Raymond, Rossiter W., 1870
Raymond, Rossiter W., 1871
Raymond, Rossiter W., 1872
Raymond, Rossiter W., 1873
Raymond, Rossiter W., 1874
Raymond, Rossiter W., 1875
Raymond, Rossiter W., 1877

Kansas
Elliott, R. S., 1872

Lake Superior region
Featherstonhaugh, George W., 1836
Foster, J. W., & Whitney, J. D., 1850a

Mexico
Bureau of the American Republics, 1904b

Sonora
Fergusson, D., 1863

Montana
Browne, J. R., 1867
Browne, J. R., 1868
Eaton, A. K.; Keyes, W. S., & DeLacy, W. W., 1869
Keyes, W. S., 1868
Mullan, John, 1863
Peale, A. C., 1872

Raymond, Rossiter W., 1869
Raymond, Rossiter W., 1870
Raymond, Rossiter W., 1871
Raymond, Rossiter W., 1872
Raymond, Rossiter W., 1873
Raymond, Rossiter W., 1874
Raymond, Rossiter W., 1875
Raymond, Rossiter W., 1877
U. S. Department of Agriculture, 1872

Nevada

Browne, J. R., 1867
Browne, J. R., 1868
Raymond, Rossiter W., 1869
Raymond, Rossiter W., 1870
Raymond, Rossiter W., 1871
Raymond, Rossiter W., 1872
Raymond, Rossiter W., 1873
Raymond, Rossiter W., 1874
Raymond, Rossiter W., 1875
Raymond, Rossiter W., 1877

New Mexico

Bureau of Immigration, 1901
Herrick, C. L., 1900
McCauley, C. A. H., 1878
Raymond, Rossiter W., 1870
Raymond, Rossiter W., 1871

Raymond, Rossiter W., 1872
Raymond, Rossiter W., 1873
Raymond, Rossiter W., 1874
Raymond, Rossiter W., 1877

North Dakota

Raymond, Rossiter W., 1877

Oklahoma

Van Vleet, A. H., 1907

Oregon

Browne, J. R., 1867
Browne, J. R., 1868
Raymond, Rossiter W., 1870
Raymond, Rossiter W., 1871
Raymond, Rossiter W., 1872
Raymond, Rossiter W., 1873
Raymond, Rossiter W., 1874
Raymond, Rossiter W., 1875
Raymond, Rossiter W., 1877

Panama

Blanchard, I., 1869
Raymond, Rossiter W., 1869

Philippines

Becker, George F., 1901
Eveland, Arthur J., 1907
McCaskey, H. D., 1905a
McCaskey, H. D., 1905b
McCaskey, H. D., 1907

Rocky Mountains

Dodge, G. M., 1868
Dodge, G. M., 1869

Mineral resources (*Cont'd.*)
Mullan, John, 1863
Peale, A. C., 1872
Raymond, Rossiter W., 1869
Raymond, Rossiter W., 1870
Raymond, Rossiter W., 1871
Raymond, Rossiter W., 1872
Raymond, Rossiter W., 1873
Raymond, Rossiter W., 1874
Raymond, Rossiter W., 1875
Raymond, Rossiter W., 1877
Van Lennep, D., 1868

South Dakota
Jenny, Walter P., 1876
Raymond, Rossiter W., 1877

Trinidad
Wall, G. P., & Sawkins, Joseph, 1857

United States
Featherstonhaugh, George W., 1833
Pumpelly, Raphael, 1886
U. S. Department of the Interior, 1866

United States, eastern
Featherstonhaugh, George W., 1836

United States, western
Browne, J. R., 1867
Browne, J. R., 1868

Utah
Browne, J. R., 1867
Browne, J. R., 1868

Peale, A. C., 1872
Raymond, Rossiter W., 1869
Raymond, Rossiter W., 1870
Raymond, Rossiter W., 1871
Raymond, Rossiter W., 1872
Raymond, Rossiter W., 1873
Raymond, Rossiter W., 1874
Raymond, Rossiter W., 1875
Raymond, Rossiter W., 1877

Venezuela
Goiticoa, N. V., 1904

Washington
Browne, J. R., 1867
Browne, J. R., 1868
Mullan, John, 1863

Wisconsin
Featherstonhaugh, George W., 1836

Wyoming
Darton, Nelson H., 1906
Peale, A. C., 1872
Raymond, Rossiter W., 1870
Raymond, Rossiter W., 1871
Raymond, Rossiter W., 1872
Raymond, Rossiter W., 1873
Raymond, Rossiter W., 1877

Mineral waters. See under **Mineralogy**; see also **Springs** and **Thermal springs** under **Geomorphology**.

Mineralogy. See also **Crystalography**; see also **Mineralogy** under area terms.
  Bibliography
  1885
  Dana, Edward S., 1886
  1886
  Dana, Edward S., 1889
  California
  Blake, William P., 1867
  Browne, J. R., 1867
  Chile
  Smith, J. Lawrence, 1855
  Classification
  Tassin, Wirt, 1899b
  Collections
   Land Office
   Roessler, A. R.; Wilson, Joseph S., & Goodwin, J. R., 1871
   Non-metallic minerals, U. S. National Museum
   Merrill, George P., 1901
   U. S. National Museum
   Tassin, Wirt, 1897
   Tassin, Wirt, 1899a
   Tassin, Wirt, 1899b
   Tassin, Wirt, 1902a
  Colorado
  Loew, Oscar, 1875
  Crystal growth
  Judd, John W., 1893
  Liveing, G. D., 1893
  Crystallography
  Brezina, Aristides, 1873

Diamonds
  Origin
  Williams, Gardner F., 1906
  Education
  Roessler, A. R.; Wilson, Joseph S., & Goodwin, J. R., 1871
  Collections
  Hughes, George W., et al, 1837
  Great Plains
  Hitchock, Edward, 1853
  Shepard, Charles U., 1853
  History
  Anderson, Martin B., 1871
  Cuvier, Georges, 1861
  Elie de Beaumont, L., 1871
  1881
  Hawes, George W., 1881b
  1882
  Dana, Edward S., 1884
  1883
  Dana, Edward S., 1885a
  1884
  Dana, Edward S., 1885b
  1885
  Dana, Edward S., 1886
  1886
  Dana, Edward S., 1889
  1887
  Dana, Edward S., 1890
  1888
  Dana, Edward S., 1890
  Memoirs, Dewey, C.
  Anderson, Martin B., 1871
  Idaho
  Peale, A. C., 1872

Mineralogy *(Cont'd.)*

Identification techniques
Egleston, T., 1873

Lake Superior region
Burt, William A., & Hubbard, Bela, 1849
Locke, John, 1849

Memoirs
Bravais, A.
Elie de Beaumont, L., 1871
Hauy, R. J.
Cuvier, Georges, 1861

Michigan
Burt, William A., & Hubbard, Bela, 1849

Mineral descriptions
Cullinan diamond
Match, F. H., & Corstorphine, G. S., 1906

Mineral waters
Arkansas
Haywood, J. K., 1902
California
Loew, Oscar, 1876a
Chile
Gilliss, J. M., 1855
Smith, J. Lawrence, 1855
Colorado
Smart, Charles, 1879
Japan
Jones, George, 1856a
Wyoming
Heizmann, C. L., 1874
Montana
Peale, A. C., 1872

New Mexico
Bailey, J. W., 1848
Loew, Oscar, 1875

Nicaragua
Endlich, Frederic M., 1874

Oklahoma
Hitchock, Edward, 1853
Shepard, Charles U., 1852

Optical properties
Brezina, Aristides, 1873

Physical properties
Tassin, Wirt, 1899a

Relation to geology
Lapworth, Charles, 1904

Rocky Mountains
Peale, A. C., 1872

Texas
Hitchock, Edward, 1853
Shepard, Charles U., 1853

Utah
Peale, A. C., 1872

Wyoming
Peale, A. C., 1872

Mining geology

California
King, T. Butler, 1850
Waldeyer, Charles, 1873

Education
Lyman, Benjamin S., 1869
Raymond, Rossiter W., 1869
Ward, Willard P., 1869

Berlin
Ward, Willard P., 1869

Clausthal, Frieberg, Paris, Prussia, United States
Raymond, Rossiter W., 1869

Freiberg
Lyman, Benjamin S., 1869

History
Browne, J. R., 1867
Taylor, James W., 1867
Taylor, James W., 1868

United States
Raymond, Rossiter W., 1877

Lake Superior region
Barnes, George O., 1849a
Barnes, George O., 1849b
Burt, William A., & Hubbard, Bela, 1849
Dickenson, George J., 1849
Foster, J. W., 1849a
Foster, J. W., & Whitney, J. D., 1849
Gibbs, William P., 1849
Jackson, Charles T., 1849
McIntyre, James, 1849

Methods
Browne, J. R., 1867
Raymond, Rossiter W., 1870
Taylor, James W., 1867
Taylor, James W., 1868

California, Nevada
Raymond, Rossiter W., 1873

Michigan
Burt, William A., & Hubbard, Bela, 1849
Foster, J. W., & Whitney, J. D., 1849

Mining law

Foreign, History of
Raymond, Rossiter W., 1869

United States
Raymond, Rossiter W., 1872

Nevada
Church, John A., 1877
Church, John A., 1878

Minnesota

Areal geology
Foster, J. W., & Whitney, J. D., 1851
Owen, David D., 1848b
Pope, John, 1850

Economic geology

Gold
Taylor, James W., 1867
Taylor, James W., 1868

Iron
Foster, J. W., & Whitney, J. D., 1851
Pumpelly, Raphael, 1886

Engineering geology

Bridges
Warren, G. K., 1878

Rivers
Warren, G. K., 1875b

Geological exploration
Pope, John, 1850

Geomorphology

Fluvial features
Warren, G. K., 1875b
Warren, G. K., 1878

Glacial features
Owen, David D., 1848b

Minnesota (Cont'd.)

Landform description
Hayden, F. V., 1873
Thomas, Cyrus, 1873
Warren, G. K., 1875b

Soils, Blue Earth County
Bennett, Hugh H., & Hurst, Lewis A., 1908

Soils, Carlton
Geib, W. J., & Jones, Grove B., 1907

Soils, Crookston
Mangum, A. W., & Schroeder, F. C., 1908

Soils, Marshall
Wilder, Henry J., 1904

Soils, Superior
Caine, Thomas A., & Lyman, W. S., 1905b

Terraces
Desor, E., 1851

Glacial geology
Drift, Glacial lakes
Desor, E., 1851

Paleontology
Invertebrata
Hall, James, 1851a
Owen, David D., 1848b

Stratigraphy
Hall, James, 1851b

Mississippi
Areal geology
Eckel, Edwin C., & Crider, A. F., 1905
Hilgard, Eugene W., 1884d

Economic geology
Limestone
Eckel, Edwin C., & Crider, A. F., 1905
U. S. Department of Agriculture, 1870

Geomorphology
Fluvial features, Landform description, Loess
Little, George, 1882

Landform description
Hilgard, Eugene W., 1884d

Soils
Hilgard, Eugene W., 1884d

Soils, Biloxi
Hearn, W. Edward, & Carr, M. E., 1905

Soils, Crystalsprings
Burgess, James L., & Tharp, W. E., 1907

Soils, Jackson
Martin, J. O., & Ayrs, O. L., 1905

Soils, McNeill
Smith, William G., & Carter, William T., Jr., 1904b

Soils, Montgomery County
Caine, Thomas A., & Schroeder, Frank C., 1908

Soils, Pontotoc County
Bennett, Frank, Jr., & Winston, R. A., 1908

Soils, Smedes
Smith, William G., & Carter, William T., Jr., 1903

Soils, Yazoo
Bonsteel, Jay A., et al, 1902b

Stratigraphy
Quaternary
Little, George, 1882

Mississippi Valley. See also names of appropriate states.
Areal geology, Geological exploration
Nicollet, I. N., 1843
Engineering geology
Bridges
Warren, G. K., 1878
Floods
Cowden, John, 1878
Floods, Rivers
Warren, G. K., 1875a
Geomorphology
Fluvial features
Cowden, John, 1878
Little, George, 1882
Warren, G. K., 1875a
Warren, G. K., 1878
Landform description
Hilgard, Eugene W., 1884a
Landform description, Loess
Little, George, 1882
Paleontology
Invertebrata
Nicolet, I. N., 1843
Stratigraphy
Quaternary
Little, George, 1882

Missouri
Areal geology
Featherstonhaugh, George W., 1835
Loughridge, R. H., 1884a
Nicollet, I. N., 1843
Earthquakes
1811
Dudley, Timothy, 1859
Economic geology
Coal
Jamison, J., 1853
Iron
Pumpelly, Raphael, 1886
Engineering geology
Bridges
Warren, G. K., 1878
Geological exploration
Featherstonhaugh, George W., 1835
Nicollet, I. N., 1843
Geomorphology
Fluvial features
Warren, G. K., 1878
Landform description
Loughridge, R. H., 1884a
Soils
Loughridge, R. H., 1884a
Soils, Crawford County
Hearn, W. Edward, & Mann, Charles J., 1907a
Soils, Howell County
Fippin, Elmer O., 1903b
Soils, O'Fallon
Fippin, Elmer O., & Drake, J. A., 1905b
Soils, Putnam County
Mann, Charles J., & Tharp, W. E., 1908
Soils, Saline County
Carr, M. Earl, & Belden, H. L., 1905

Missouri (Cont'd.)

Soils, Scotland County
Hearn, W. Edward, & Mann, Charles J., 1907b

Soils, Shelby County
Avon-Burke, R. T., & Ruhlen, La Mott, 1904b

Soils, Webster County
Drake, J. A., & Strahorn, A. T., 1905

Paleontology

Invertebrata
Hayden, F. V., 1883a
Nicollet, I. N., 1843
White, Charles A., 1883

Mollusca

Carpenter, Philip P., 1861

Montana

Areal geology
Dana, Edward S., & Grinnell, George B., 1876
Ludlow, William, 1876

Economic geology

Copper, Gold, Silver
Taylor, James W., 1867
Taylor, James W., 1868

Mineral resources
Browne, J. R., 1867
Browne, J. R., 1868
Eaton, A. K.; Keyes, W. S., & DeLacy, W. W., 1869
Keyes, W. S., 1868
Mullan, John, 1863
Peale, A. C., 1872
Raymond, Rossiter W., 1869
Raymond, Rossiter W., 1870
Raymond, Rossiter W., 1871
Raymond, Rossiter W., 1872
Raymond, Rossiter W., 1873
Raymond, Rossiter W., 1874
Raymond, Rossiter W., 1875
Raymond, Rossiter W., 1877
U. S. Department of Agriculture, 1872

Engineering geology

Railroads
Powell, John W., 1886b

Railroads, Roads
Mullan, John, 1863

Geological exploration
Barlow, J. W., 1872
Doane, Gustavus C., 1871
Hayden, F. V., 1872a
Hayden, F. V., 1873
Hayden, F. V., 1878b
Mullan, John, 1863
Peale, A. C., 1873
Wheeler, George M., 1874
Wheeler, George M., 1875
Wheeler, George M., 1876
Wheeler, George M., 1877
Wheeler, George M., 1878
Wheeler, George M., 1879a

Geomorphology

Landform description
Dana, Edward S., & Grinnell, George B., 1876
Doane, Gustavus C., 1871
Ludlow, William, 1876
Matthes, Francois E., 1905
Thomas, Cyrus, 1872
U. S. Department of Agriculture, 1872

Soils, Billings
Jensen, Charles A., & Neill, N. P., 1903b
Soils, Gallatin Valley
Lapham, Macy H., & Ely, Charles W., 1907
Glacial geology
Exploration
Chaney, L. W., 1905
Glacial features
Matthes, Francois E., 1905
Mineralogy
Peale, A. C., 1872
Paleontology
Invertebrata
Meek, F. B., 1873
Whitfield, R. P., 1876
Paleobotany
Lesquereux, L., 1872a
Vertebrata
Cope, E. D., 1884

Morocco
Areal geology
Fischer, Theobald, 1905
Geomorphology
Landform description
Fischer, Theobald, 1905

Mountain building. See Orogeny; see Tectonics.

Mountains. See under Tectonics.

Mudlumps. See under Sedimentary structures; see Special features under Deltas.

Museums
Library of Congress
Mineralogy collection
Hughes, George W., et al, 1837
Smithsonian Institution
History
Rhees, William J., 1901
U. S. National Museum
Building stone collection
Merrill, George P., 1889
Gem collection
Kunz, George F., 1889
Tassin, Wirt, 1902a
History
True, Frederick W., 1898
Meteorite collection
Clarke, F. W., 1889
Tassin, Wirt, 1902b
Mineral collection
Tassin, Wirt, 1897
Tassin, Wirt, 1899a
Tassin, Wirt, 1899b
Non-metallic minerals
Merrill, George P., 1901
Origin
Goode, George B., 1901
Ornamental stone collection
Merrill, George P., 1889
Rock and mineral collection
Merrill, George P., 1891

N

Nebraska
Areal geology, Geological exploration
Engelmann, Henry, 1858
Gregory, J. W., 1892

Nebraska (Cont'd.)

Hayden, F. V., 1858
Hicks, L. E., 1892
Warren, G. K., 1856

Geological exploration
Engelmann, Henry, 1858
Hayden, F. V., 1871
Hayden, F. V., 1872b
Owen, David D., 1852
Warren, G. K., 1858
Wheeler, George M., 1875
Wheeler, George M., 1878
Wheeler, George M., 1879a

Geomorphology
Landform description
Hayden, F. V., 1873
Thomas, Cyrus, 1873

Soils, Grand Island
Hearn, W. Edward, &
 Burgess, James L., 1904a

Soils, Kearney
Martin, J. O., & Sweet, A.
 T., 1905

Soils, Lancaster County
Burgess, James L., &
 Worthen, H. L., 1908

Soils, Sarpy County
Kocher, A. E., & Hurst,
 Lewis A., 1907b

Soils, Stanton
Hearn, W. Edward, 1904

Hydrogeology
Gregory, J. W., 1892
Hicks, L. E., 1892
U. S. Department of Agriculture, 1890

Paleontology
Hayden, F. V., 1858

Invertebrata
Meek, F. B., 1871
Meek, F. B., 1872c
Shumard, B. F., 1858

Mammalia
Leidy, Joseph, 1852

Paleobotany
Lesquereux, L., 1872b

Pisces, Vertebrata
St. John, Orestes H., 1871

Vertebrata
Leidy, Joseph, 1852

Nevada

Areal geology
Blatchly, A., 1867
Ives, Joseph C., 1861
Newberry, J. S., 1861
Sitgreaves, L., 1853

Economic geology
Coal
Lesquereux, L., 1873

Gold, Silver
Blatchly, A., 1867
Edmunds, J. M., et al, 1863

Methods
Wright, H. G.; Foster, J. G.,
 & Newcomb, Wesley,
 1872

Mineral resources
Browne, J. R., 1867
Browne, J. R., 1868
Raymond, Rossiter W.,
 1869
Raymond, Rossiter W.,
 1870
Raymond, Rossiter W.,
 1871
Raymond, Rossiter W.,
 1872

Raymond, Rossiter W., 1873
Raymond, Rossiter W., 1874
Raymond, Rossiter W., 1875
Raymond, Rossiter W., 1877

Practice
Church, John A., 1877
Church, John A., 1878

Engineering geology

Tunnels
Wright, H. G.; Foster, J. G., & Newcomb, Wesley, 1872

Geological exploration
Conkling, A. R., 1877a
Conkling, A. R., 1878
Gilbert, G. K., 1872
Ives, Joseph C., 1861
King, Clarence, 1871
King, Clarence, 1873
King, Clarence, 1874
King, Clarence, 1876
King, Clarence, 1877
Lockwood, Daniel W., 1872
Lyle, D. A., 1872
Newberry, J. S., 1861
Powell, John W., 1877
Sitgreaves, L., 1853
Wheeler, George M., 1872a
Wheeler, George M., 1872b
Wheeler, George M., 1873
Wheeler, George M., 1874
Wheeler, George M., 1875
Wheeler, George M., 1876
Wheeler, George M., 1877
Wheeler, George M., 1878
Wheeler, George M., 1879a

Geomorphology

Geysers
Peale, A. C., 1883

Landform description
Powell, John W., 1878b
U. S. Department of Agriculture, 1871

Paleontology

Invertebrata
Newberry, J. S., 1861

Vertebrata
Cope, E. D., 1884

New Hampshire

Economic geology

Gold
Taylor, James W., 1867

Iron
Pumpelly, Raphael, 1886

Geomorphology

Soils, Merrimack County
Mooney, Charles N.; Westover, H. L., & Bennett, Frank, 1908

New Jersey

Economic geology

Iron
Pumpelly, Raphael, 1886

Geomorphology

Deltas
Mitchell, Henry, 1887b

Fluvial features
Marinden, Henry L., 1882
Marinden, Henry L., 1883
Marinden, Henry L., 1885
Mitchell, Henry, 1881
Mitchell, Henry, 1884
Mitchell, Henry, 1887b
Mitchell, Henry, 1889a

Shore features
Mitchell, Henry, 1884
Mitchell, Henry, 1889a

New Jersey (Cont'd.)

Soils, Salem
Bonsteel, Jay A., & Taylor, F. W., 1902

Soils, Trenton
Avon-Burke, R. T., & Wilder, Henry J., 1903

Marine geology

Continental shelf
Lindenkohl, A., 1885

Shore features
Mitchell, Henry, 1889a

Surveys

State Geological Survey
Kümmel, Henry B., 1905

New Mexico

Areal geology
Abert, J. W., 1846
Antisell, Thomas, 1855
Beale, Edward F., 1858
Beale, Edward F., & Engle, F. E., 1860
Bureau of Immigration, 1901
Conkling, A. R., 1876
Conrad, T. A., 1857
Cope, E. D., 1875
Emmons, S. F., 1893
Emory, William H., 1855
Emory, William H., 1857
Governor of New Mexico, 1905
Hall, James, 1857
Herrick, C. L., 1900
Herrick, C. L., 1901
Hill, Robert T., 1892
Johnston, A. R., 1858
Keyes, Charles R., 1903
Loew, Oscar, 1875
Parke, John G., 1855a
Parke, John G., 1855b

Parry, C. C., & Schott, Arthur, 1857
Pope, John, 1860
Simpson, J. H., 1850b
Simpson, J. H., & Marcy, R. B., 1850
Sitgreaves, L., 1853
Wislezenus, A., 1848

Economic geology

Coal
Lesquereux, L., 1873
Sheridan, Jo E., 1903
Sheridan, Jo E., 1904

Copper, Gold, Lead, Silver
Taylor, James W., 1867
Taylor, James W., 1868

Mineral resources
Bureau of Immigration, 1901
Herrick, C. L., 1900
McCauley, C. A. H., 1878
Raymond, Rossiter W., 1870
Raymond, Rossiter W., 1871
Raymond, Rossiter W., 1872
Raymond, Rossiter W., 1873
Raymond, Rossiter W., 1874
Raymond, Rossiter W., 1877

Petroleum
Herrick, C. L., 1901

Engineering geology

Railroads
Parke, John G., 1855a
Parke, John G., 1855b

Geological exploration
Abert, J. W., 1846
Abert, J. W., 1848

Index 235

Antisell, Thomas, 1855
Beale, Edward F., 1858
Beale, Edward F., & Engle, F. E., 1860
Conkling, A. R., 1876
Conkling, A. R., 1877b
Emory, William H., 1855
Emory, William H., 1857
Ives, Joseph C., 1858
Johnston, A. R., 1858
McCauley, C. A. H., 1878
Marcy, Randolph B., 1850
Parke, John G., 1855a
Parke, John G., 1855b
Parry, C. C., & Schott, Arthur, 1857
Ruffner, E. H., 1874
Simpson, J. H., 1850a
Simpson, J. H., 1850b
Simpson, J. H., & Marcy, R. B., 1850
Sitgreaves, L., 1853
Stevenson, J. T., 1879
Wheeler, George M., 1873
Wheeler, George M., 1874
Wheeler, George M., 1875
Wheeler, George M., 1876
Wheeler, George M., 1877
Wheeler, George M., 1878
Wheeler, George M., 1879a
Wislezenus, A., 1848

Geomorphology

Landform description

McCauley, C. A. H., 1878
Powell, John W., 1878b

Soils, Pecos Valley

Means, Thomas H., & Gardner, Frank D., 1900

Hydrogeology

Hill, Robert T., 1892
Pope, John, 1860
U. S. Department of Agriculture, 1890

Mineralogy

Bailey, J. W., 1848
Loew, Oscar, 1875

Paleontology

Abert, J. W., 1848

Invertebrata

Bailey, J. W., 1848
Conrad, T. A., 1857
Cope, E. D., 1875
Hall, James, 1857

Vertebrata

Cope, E. D., 1874
Cope, E. D., 1875
Cope, E. D., 1884

Petrology

Conkling, A. R., 1877c

Stratigraphy

Keyes, Charles R., 1903

New York

Areal geology

Emmons, S. F., 1893

Economic geology

Iron

Pumpelly, Raphael, 1886

Geomorphology

Fluvial features

Tarr, R. S., 1905

Shore features, beaches

Bache, A. D., 1856b

Soils, Auburn

Lapham, J. E., & Bennett, Hugh H., 1905

Soils, Bigflats

Mesmer, Louis, & Hearn, W. E., 1903

New York (Cont'd.)

Soils, Binghamton
Fippin, Elmer O., & Carter, William T., Jr., 1907

Soils, Long Island
Bonsteel, Jay A., et al, 1904

Soils, Lyons
Hearn, W. Edward, 1903

Soils, Madison County
Carr, M. Earl; Griffin, A. M., & Lee, Ora, Jr., 1908

Soils, Niagara County
Fippin, Elmer O., & Mann, C. W., 1908

Soils, Syracuse
Bonsteel, F. E.; Carter, William T., Jr., & Ayres, O. L., 1904

Soils, Tompkins County
Bonsteel, Jay A.; Fippin, Elmer O., & Carter, William T., Jr., 1907

Soils, Vergennes
Wilder, Henry J., & Belden, H. L., 1905b

Soils, Westfield
Avon-Burke, R. T., & Marean, Herbert W., 1902b

Glacial geology
Glacial features
Tarr, R. S., 1905

Pleistocene lakes
Colemen, A. P., 1905

Marine geology
Continental shelf
Lindenkohl, A., 1885

Shore features
Bache, A. D., 1856b

New Zealand

Geomorphology
Geysers
Peale, A. C., 1883
Weed, Walter H., 1893

Newfoundland

Areal geology
Lieber, Oscar M., 1861

Geomorphology
Landform description
Lieber, Oscar M., 1861

Nicaragua

Areal geology
Davis, C. H., 1866
Hatfield, Chester, 1874
Whitfield, Benjamin, 1874

Earthquakes
Dutton, C. E., 1902
Jones, James O., 1903
Wheeler, E. S., 1902

Economic geology
Construction materials
Powell, John W., 1886a

Engineering geology
Canals
Dutton, C. E., 1902
Hatfield, Chester, 1874
Wheeler, E. S., 1902

Canals, Railroads
Davis, C. H., 1866

Geological exploration
Hatfield, Chester, 1874
Whitfield, Benjamin, 1874

Geomorphology
Fluvial features
Davis, Arthur P., 1903

Landform description
Hatfield, Chester, 1874
Whitfield, Benjamin, 1874
Shore features
Mitchell, Henry, 1877
Hydrogeology
Davis, Arthur P., 1903
Marine geology
Shore features
Mitchell, Henry, 1877
Mineralogy, Petrology
Endlich, Frederic M., 1874
Structural geology
Orogeny and uplifts
Sapper, Karl, 1905
Volcanology
Chamberlain, P. W., 1903
Dutton, C. E., 1902
Jones, James O., 1903
Sapper, Karl, 1905

North America
Geomorphology
Landform description
Froebel, Julius, 1855
Glacial geology
Animal and plant distribution
Adams, Charles C., 1905
Plant distribution
Anderson, Richard J., 1905
Harshberger, J. W., 1905
History of geology
1886
Darton, Nelson H., 1889

Paleoclimatology
Plant distribution, Quaternary
Harshberger, J. W., 1905
Paleontology
Bibliography, 1885
Marcou, John B., 1886
Bibliography, 1886
Marcou, John B., 1889
History, 1887-1888
Williams, Henry S., 1890
Invertebrata, Crustacea
Packard, A. S., 1886b
Invertebrata, 1884
Marcou, John B., 1885
Invertebrata, 1885
Marcou, John B., 1886
Invertebrata, 1886
Marcou, John B., 1889
Paleobotany, Quaternary
Anderson, Richard J., 1905

North Carolina
Areal geology
Keith, Arthur, 1902
Kerr, W. C., 1884a
Earthquakes
1874
du Pre, Warren, 1875
Economic geology
Coal, Iron
Laidley, T. T. S., 1856
Wilkes, Charles; Hunt, H., & Martin, D. B., 1859
Gold
Taylor, James W., 1867
Iron
Pumpelly, Raphael, 1886

North Carolina (Cont'd.)

Engineering geology

Shorelines
U. S. Army Corps of Engineers, 1905

Geomorphology

Landform description
Keith, Arthur, 1902
Kerr, W. C., 1884a

Shore features
U. S. Army Corps of Engineers, 1905

Soils
Kerr, W. C., 1884a

Soils, Alamance County
Coffey, George N., & Hearn, W. Edward, 1902a

Soils, Asheville
Lapham, J. E., & Meeker, F. N., 1904

Soils, Cary
Coffey, George N., & Hearn, W. Edward, 1902b

Soils, Chowan County
Hearn, W. Edward, & MacNider, G. M., 1908a

Soils, Craven
Smith, William G., & Coffey, George N., 1904

Soils, Duplin County
Root, Aldert S., & Hurst, Lewis A., 1907

Soils, Greeneville
Mooney, Charles N., & Ayrs, O. L., 1905a

Soils, Hickory
Caine, Thomas A., 1903

Soils, Mount Mitchell
Caine, Thomas A., & Mangum, A. W., 1903

Soils, New Hanover County
Drake, J. A., & Belden, H. L., 1908

Soils, Pasquotank County
Lapham, J. E., & Lyman, W. S., 1907

Soils, Perquimans County
Lapham, J. E., & Lyman, W. S., 1907

Soils, Raleigh to Newbern
Smith, William G., 1901

Soils, Statesville
Dorsey, Clarence W., et al, 1902

Soils, Transylvania County
Hearn, W. Edward, & MacNider, G. M., 1908b

Marine geology

Shore features
U. S. Army Corps of Engineers, 1905

North Dakota

Economic geology

Mineral resources
Raymond, Rossiter W., 1877

Geomorphology

Landform description
Hayden, F. V., 1873
Thomas, Cyrus, 1873

Soils, Cando
Fippin, Elmer O., & Burgess, James L., 1905

Soils, Carrington
Kocher, A. E., & Hurst, Lewis A., 1907a

Soils, Fargo
Caine, Thomas A., 1904

Soils, Grand Forks
Jensen, Charles A., & Neill, N. P., 1903a

Soils, Jamestown
Caine, Thomas A., & Kocher, A. E., 1904a

Soils, Ransom County
Ely, Charles W.; Willard, Rex E., & Weaver, J. T., 1908

Soils, Williston
Rice, Thomas D., 1908

Hydrogeology
U. S. Department of Agriculture, 1890

Northwest Territories

Geomorphology
  Landform description
  Hall, Charles F., 1879

Glacial geology
  Ice
  Hall, Charles F., 1879

Paleontology
  Emerson, Benjamin K., 1879

Petrology
  Rock description
  Emerson, Benjamin K., 1879

Norway

Surveys
  Geological and topographical surveys
  Comstock, C. B., 1876
  Land and marine surveys
  Wheeler, George M., 1885

Nova Scotia

Economic geology
  Coal, Gold
  Hamilton, Pierce S., 1868
  Gold
  Taylor, James W., 1867
  Taylor, James W., 1868

Geomorphology
  Shore features
  Mitchell, Henry, 1880

Glacial geology
Isostatic rebound
  Mitchell, Henry, 1880

Marine geology
  Shore features
  Mitchell, Henry, 1880

O

Ocean basins, see under **Marine geology**.

Oceanography, see also **Marine geology**.
  Abyssal fauna
    Origin
    Ortmann, A. E., 1905
    Challenger expedition
    Daubrée, A., 1894

Oceanography (*Cont'd.*)
History
Henry, Joseph, 1871
Instruments
Craven, T. A., 1855
Maps
Bathymetric
Thoulet, J., 1905
Memoirs
Bache, A. D.
Henry, Joseph, 1871

Ohio
Economic geology
Coal
Jamison, J., 1853
Gas, natural
Orton, Edward, 1893
Iron
Pumpelly, Raphael, 1886
Engineering geology
Roads
Macomb, Alexander; Knight, Jonathan, & Wevers, C. W., 1826
Geomorphology
Soils, Ashtabula
Martin, J. O., & Carr, E. P., 1904a
Soils, Cleveland
Lapham, J. E., & Mooney, Charles N., 1907a
Soils, Columbus
Smith, William G., 1903a
Soils, Coshocton County
Rice, Thomas D., & Geib, W. J., 1905a

Soils, Meigs County
Meeker, F. N., & Tailby, G. W., Jr., 1908
Soils, Montgomery County
Dorsey, Clarence W., & Coffey, George N., 1901
Soils, Toledo
Smith, William G., 1903b
Soils, Westerville
Lapham, J. E., & Mooney, Charles A., 1907b
Soils, Wooster
Caine, Thomas A., & Lyman, W. S., 1905c
Structural geology
Rock pressure
Orton, Edward, 1893

Oklahoma
Areal geology
Gregory, J. W., 1892
Loughridge, R. H., 1884d
Marcy, Randolph B., 1853
Shumard, B. F., 1853b
Van Vleet, A. H., 1904
Van Vleet, A. H., 1907
Economic geology
Coal, Copper, Gypsum
Marcy, Randolph B., 1853
Gold
Bain, H. Foster, 1904
Gypsum, Salt
Van Vleet, A. H., 1904
Mineral resources
Van Vleet, A. H., 1907
Geological exploration
Marcy, Randolph B., 1853
Ruffner, E. H., 1877a

Shumard, B. F., 1853b
Wheelock, T. B., 1834
Geomorphology
Landform description
Loughridge, R. H., 1884d
Soils
Loughridge, R. H., 1884d
Soils, Tishomingo
Rice, Thomas D., & Ayrs, Orla L., 1908
Hydrogeology
Gregory, J. W., 1892
Hinton, Richard J., 1892
Mineralogy, Petrology
Hitchock, Edward, 1853
Shepard, Charles U., 1853
Paleontology
Invertebrata
Shumard, B. F., 1853a
Vertebrata, Mammalia
Holmes, William H., 1903a

Ontario

Geomorphology
Landform description
Wilson, A. W. G., 1905
Glacial geology
Isostatic rebound
Bell, Robert, 1898
Pleistocene lakes
Coleman, A. P., 1905

Oregon

Areal geology
Abbot, Henry L., 1855
Cross, Osborne, 1850
Fremont, John C., 1845
Fremont, John C., 1848

Newberry, J. S., 1855
Stevens, I. I., 1855a
Stevens, I. I., 1855b
Economic geology
Coal, Iron
Gabb, W. M., 1867
Limestone
Frazer, John F., 1850
Mineral resources
Browne, J. R., 1867
Browne, J. R., 1868
Raymond, Rossiter W., 1870
Raymond, Rossiter W., 1871
Raymond, Rossiter W., 1872
Raymond, Rossiter W., 1873
Raymond, Rossiter W., 1874
Raymond, Rossiter W., 1875
Raymond, Rossiter W., 1877
Engineering geology
Railroads
Abbot, Henry L., 1855
Stevens, I. I., 1855a
Stevens, I. I., 1855b
Geological exploration
Abbot, Henry L., 1855
Cross, Osborne, 1850
Fremont, John C., 1845
Fremont, John C., 1848
Newberry, J. S., 1855
Stevens, I. I., 1855a
Stevens, I. I., 1855b
Wheeler, George M., 1877
Wheeler, George M., 1878
Wheeler, George M., 1879a

Oregon (Cont'd.)

Geomorphology
Fluvial features, Shore features
  Langfitt, W. C., 1903
Lacustrine features, Crater Lake
  Diller, J. S., 1898
Landform description
  Hergesheimer, E., 1883
Soils, Baker City
  Jensen, Charles A., & Mackie, W. W., 1904
Soils, Salem
  Jensen, Charles A., 1904
Thermal springs
  Dornbach, L. M., 1855

Glacial geology
Mt. Hood
  Reid, Harry F., 1905a

Marine geology
Shore features
  Langfitt, W. C., 1903

Mineralogy
Mineral waters
  Dornbach, L. M., 1855

Paleontology
Diatoms
  Bailey, J. W., 1845
Invertebrata
  Conrad, T. A., 1855b
  Hall, James, 1845a
Vertebrata
  Cope, E. D., 1884

Stratigraphy
  Hall, James, 1845b

Volcanology
Crater Lake
  Diller, J. S., 1898

Orogeny. See also **Tectonics.**
Causes
  Sollas, W. J., 1901
Lake Superior region
  Foster, J. W., & Whitney, J. D., 1851
Central America
  Sapper, Karl, 1905
Laramide
  Hayden, F. V., 1878d

P

Pacific islands
Areal geology
  Fahs, C. F., 1856
  Perry, M. C., 1856
  Taylor, Bayard, 1856

Pacific Ocean
Marine geology
  Bottom features
    Clover, Richardson, 1892
  Coral atolls
    Sollas, W. J., 1899
  Sediments
    Clover, Richardson, 1892

Paleobotany. See also **Paleontology**; see also **Diatoms; Pteridospermophyta.**
Arizona
  Triassic
    Lacey, John F., 1904
    Lacey, John F., 1906
    Ward, Lester F., 1901

California
  Bailey, J. W., 1855a
  Schaeffer, George C., 1855
Carboniferous
  LeConte, Joseph, 1858
Colorado
  Tertiary
  Lesquereux, L., 1873
Geographic distribution
  Germany, Quaternary
  Drude, Oscar, 1905
  Ireland and North America
  Anderson, Richard J., 1905
  North America, Quaternary
  Harshberger, J. W., 1905
Great Basin, Great Plains, Rocky Mountains
  Lesquereux, L., 1872a
History
  Dana, James D., 1890
  Farlow, William G., 1890
Kansas, Nebraska
  Lesquereux, L., 1872b
Memoirs
  Gray, Asa
  Dana, James D., 1890
  Farlow, William G., 1890
Montana, Utah, Wyoming
  Lesquereux, L., 1872a
Nevada, New Mexico, Utah, Wyoming
  Lesquereux, L., 1873
Rocky Mountains
  Lesquereux, L., 1873

United States
  Pteridospermophyta
  White, David, 1905
Paleoclimatology
  Carboniferous
  LeConte, Joseph, 1858
  Cycles
  Blytt, A., 1890
  Quaternary
  Europe
  Geikie, James, 1899
Paleogeography
  Pilar, George, 1877
  Ireland
  Anderson, Richard J., 1905
Paleontology. See also the names of fossil groups.
  Abyssal fauna
  Origin
  Ortmann, A. E., 1905
  Arctic
  Emerson, Benjamin K., 1879
  Arizona
  Invertebrata
  Conrad, T. A., 1857
  Hall, James, 1857
  Newberry, J. S., 1861
  Paleobotany
  Lacey, John F., 1904
  Lacey, John F., 1906
  Ward, Lester F., 1901
  Bibliography
  1887-1888
  Williams, Henry S., 1890

Paleontology (Cont'd.)

California
Agassiz, Louis, 1855

Diatoms
Bailey, J. W., 1855b

Invertebrata
Conrad, T. A., 1855a
Conrad, T. A., 1855b
Conrad, T. A., 1855d
Conrad, T. A., 1857
Hall, James, 1845a
Hall, James, 1855
Hall, James, 1857
Newberry, J. S., 1861
Schiel, James, 1855b

Man, fossil
Holmes, William H., 1901

Paleobotany
Bailey, J. W., 1855a
Schaeffer, George C., 1855

Cambrian
Brooks, William K., 1896

Chile
Invertebrata
Conrad, T. A., 1855c

Vertebrata, Mammalia
Wyman, Jeffries, 1855

Colorado
Arachnida, Insecta
Scudder, Samuel H., 1883

Arthropoda
Scudder, Samuel H., 1885

Invertebrata
Hayden, F. V., 1883a
Meek, F. B., 1872c
Scudder, Samuel H., 1883
Scudder, Samuel H., 1885
White, Charles A., 1883

Paleobotany
Lesquereux, L., 1873

Vertebrata
Cope, E. D., 1884

Colorado Plateau
Invertebrata
Newberry, J. S., 1861

Diatoms
Bailey, J. W., 1845
Bailey, J. W., 1855b

Egypt
Vertebrata
Andrews, C. W., 1907

Europe, Quaternary
Rütimeyer, L., 1862

Europe
Man, fossil
Obermaier, Hugues, 1906
Suess, Edward, 1873

Fossilization
White, Charles A., 1893

Freezing
Herz, O. F., 1904

France
Man, fossil
Gaudry, Albert, 1903

Fresh water fauna
Origin
Gill, Theodore N., 1905

Great Basin
Invertebrata
Hall, James, 1852
Hall, James, 1855
Schiel, James, 1855b

Paleobotany
Lesquereux, L., 1872a
Great Plains
Invertebrata
Hall, James, 1852
Hall, James, 1855
Meek, F. B., 1872c
Schiel, James, 1855b
Shumard, B. F., 1853a
Shumard, B. F., 1858
Paleobotany
Lesquereux, L., 1872a
Great Plains, Tertiary
Vertebrata
Leidy, Joseph, 1872b
History
Arago, D. F. J., 1872
Brooks, William K., 1901
de Quatrefages, J. L. A., 1863
Favre, Ernest, 1879
Fischer, P., 1873
Fiske, John, 1901
Flourens, P., 1862
Flourens, P., 1866
Flourens, P., 1869a
Flourens, P., 1869b
Gill, Theodore N., 1896
Schuchert, Charles, 1905
1887
Williams, Henry S., 1890
1888
Williams, Henry S., 1890
Idaho
Invertebrata
Hayden, F. V., 1883a
Meek, F. B., 1873
White, Charles A., 1883
Vertebrata
Cope, E. D., 1884

Illinois
Invertebrata
Hayden, F. V., 1883a
Nicollet, I. N., 1843
White, Charles A., 1883
Indiana
Invertebrata
Hayden, F. V., 1883a
White, Charles A., 1883
Invertebrata
Prestwich, Joseph, 1876
Iowa
Invertebrata
Hayden, F. V., 1883a
Nicollet, I. N., 1843
Owen, David D., 1848b
White, Charles A., 1883
Kansas
Hayden, F. V., 1858
Invertebrata
Shumard, B. F., 1858
Man, fossil
Holmes, William H., 1903b
Paleobotany
Lesquereux, L., 1872b
Vertebrata
Cope, E. D., 1872b
Cope, E. D., 1872c
Cope, E. D., 1884
Lake Superior region
Invertebrata
Hall, James, 1851a
Locke, John, 1848
Locke, John, 1849
Owen, David D., 1848b
Life, origin
Brooks, William K., 1896
Chamberlain, T. C., 1901
Kelvin, William T., 1898

Paleontology (Cont'd.)

Louisiana
Diatoms, Foraminifera, Invertebrata
Hilgard, Eugene W., & Hopkins, F. V., 1878

Mammalia
Quaternary
Herz, O. F., 1904
Lucas, Frederick A., 1901b
Lydekker, R., 1901

Man, fossil
Anthony, M., 1904
Dubois, Eugene, 1899
Haeckel, Ernst, 1899
Suess, Edward, 1873

Memoirs
Agassiz, Louis
Favre, Ernest, 1879

Cuvier, G.
Flourens, P., 1869b

de Blainville, D.
Flourens, P., 1866

Fourier, J.
Arago, D. F. J., 1872

Huxley, T. H.
Brooks, William K., 1901
Fiske, John, 1901
Gill, Theodore N., 1896

Lartet, Edward
Fischer, P., 1873

Saint Hilaire, G.
de Quatrefages, J. L. A., 1863
Flourens, P., 1862

von Zittel, K. A.
Schuchert, Charles, 1905

Methods
Huxley, Thomas H., 1871
White, Charles A., 1893

Mexico
Invertebrata, Baja California
Newberry, J. S., 1861

Invertebrata, Baja California, Chihuahua, Coahuila, Nuevo Leon, Sonora, Tamaulipas
Conrad, T. A., 1857
Hall, James, 1857

Michigan, Minnesota
Invertebrata
Hall, James, 1851a
Locke, John, 1848

Mississippi Valley
Invertebrata
Nicollet, I. N., 1843

Missouri
Invertebrata
Hayden, F. V., 1883a
White, Charles A., 1883

Montana
Invertebrata
Meek, F. B., 1873
Whitfield, R. P., 1876

Paleobotany
Lesquereux, L., 1872a

Vertebrata
Cope, E. D., 1884

Nebraska
Hayden, F. V., 1858

Invertebrata
Meek, F. B., 1871
Meek, F. B., 1872c
Shumard, B. F., 1858

Paleobotany
Lesquereux, L., 1872b
Vertebrata
Leidy, Joseph, 1852
St. John, Orestes H., 1871
Nevada
Invertebrata
Newberry, J. S., 1861
Paleobotany
Lesquereux, L., 1873
Vertebrata
Cope, E. D., 1884
New Mexico
Invertebrata
Abert, J. W., 1848
Bailey, J. W., 1848
Conrad, T. A., 1857
Cope, E. D., 1875
Hall, James, 1857
Paleobotany
Lesquereux, L., 1873
Vertebrata
Cope, E. D., 1874
Cope, E. D., 1875
Cope, E. D., 1884
North America
Bibliography-1884, Invertebrata-1884
Marcou, John B., 1885
Bibliography-1885, Invertebrata-1885
Marcou, John B., 1886
Bibliography-1886, Invertebrata-1886
Marcou, John B., 1889
Northwest Territories
Emerson, Benjamin K., 1879

Oklahoma
Invertebrata
Shumard, B. F., 1853a
Oregon
Invertebrata
Conrad, T. A., 1855b
Hall, James, 1845a
Vertebrata
Cope, E. D., 1884
Principles
Huxley, Thomas H., 1871
Quaternary
Mammalia
Herz, O. F., 1904
Lucas, Frederick A., 1901b
Lydekker, R., 1901
Man, fossil
Anthony, Raoul, 1904
Relation to geology
Lapworth, Charles, 1904
Reptilia, Permian
Cope, E. D., 1885
Reptilia, Pterodactylia
Langley, S. P., 1902
Restorations
Lucas, Frederick A., 1901a
Elephas primigenius
Pfizenmayer, E., 1907
Rocky Mountains
Invertebrata
Hall, James, 1852
Hall, James, 1855
Meek, F. B., 1872c
Schiel, James, 1855b
Paleobotany
Lesquereux, L., 1872a
Lesquereux, L., 1873

Paleontology (Cont'd.)

Vertebrata
Cope, E. D., 1886
Leidy, Joseph, 1872b

South Dakota
Invertebrata
Grinnell, George B., 1874
Grinnell, George B., 1875
Meek, F. B., 1872c
Shumard, B. F., 1858

Texas
Invertebrata
Conrad, T. A., 1857
Hall, James, 1857
Ruffner, E. H., 1877b
Shumard, B. F., 1853a

U. S. S. R.
Mammalia, Quaternary
Herz, O. F., 1904

United States
Invertebrata, Southwest
Conrad, T. A., 1857
Hall, James, 1857

Invertebrata, Western
Hall, James, 1855

Paleobotany
White, David, 1905

Utah
Invertebrata
Hall, James, 1852
Hayden, F. V., 1883a
Meek, F. B., 1872b
Meek, F. B., 1872c
Meek, F. B., 1873
White, Charles A., 1883

Paleobotany
Lesquereux, L., 1872a
Lesquereux, L., 1873

Vertebrata
Cope, E. D., 1872d

Vertebrata, Permian
Cope, E. D., 1885

Wisconsin
Invertebrata
Hall, James, 1851a
Nicollet, I. N., 1843
Owen, David D., 1848b

Wyoming
Invertebrata
Hayden, F. V., 1883a
Meek, F. B., 1872b
Meek, F. B., 1872c
Meek, F. B., 1873
Shumard, B. F., 1858
White, Charles A., 1883

Paleobotany
Lesquereux, L., 1872a
Lesquereux, L., 1873

Vertebrata
Cope, E. D., 1872a
Cope, E. D., 1873
Cope, E. D., 1884
Leidy, Joseph, 1872a

Paleozoic. See also period names.

Lake Superior region
Correlation
Hall, James, 1851c

Panama

Areal geology
Bowditch, E. W., 1874
Carson, J. Pettigru, 1874
Davis, C. H., 1866
Selfridge, Thomas O., 1871

Earthquakes
1882
Nicaragua Canal Commission, 1906
Economic geology
Coal
Evans, John, 1860
Coal, Gold
Engle, F., 1860
Mineral resources
Blanchard, I., 1869
Raymond, Rossiter W., 1869
Engineering geology
Canals
Burr, William H., 1903
Collins, Frederick, 1879
Davis, George W., et al, 1906
Lull, E. P., 1879
Nicaragua Canal Commission, 1906
Selfridge, Thomas O., 1871
Selfridge, Thomas O., 1874
Canals, Railroads
Davis, C. H., 1866
Geological exploration
Bowditch, E. W., 1874
Carson, J. Petigru, 1874
Engle, F., 1860
Maack, G. A., 1874
Selfridge, Thomas O., 1874
Geomorphology
Fluvial features
Davis, Arthur P., 1903
Landform description
Collins, Frederick, 1879
Lull, E. P., 1879
Selfridge, Thomas O., 1874

Shore features
Mitchell, Henry, 1877
Hydrogeology
Davis, Arthur P., 1903
Marine geology
Shore features
Mitchell, Henry, 1877
Structural geology
Orogeny and Uplifts
Sapper, Karl, 1905
Volcanology
Sapper, Karl, 1905

Paraguay
Economic geology
Diamonds
Gibbon, Lardner, 1854

Pennsylvania
Economic geology
Coal
Jamison, J., 1853
Iron
Pumpelly, Raphael, 1886
Geomorphology
Fluvial features
Howard, William, 1828
Marinden, Henry L., 1882
Marinden, Henry L., 1883
Marinden, Henry L., 1885
Mitchell, Henry, 1881
Mitchell, Henry, 1884
Soils, Adams County
Wilder, Henry J., & Belden, H. L., 1905a
Soils, Chester County
Wilder, Henry J., et al, 1907a

Pennsylvania (Cont'd.)

Soils, Montgomery County
Wilder, Henry J.; Strahorn, A. T., & Geib, W. J., 1907

Soils, Lancaster
Dorsey, Clarence W., 1901

Soils, Lebanon
Smith, William G., & Bennett, Frank, Jr., 1902

Soils, Lockhaven
Martin, J. O., 1904

Peru

Areal geology
Gibbon, Lardner, 1854

Earthquakes
1868
Campbell, John V., 1871

Economic geology
Gold, Mercury, Silver
Gibbon, Lardner, 1854

Glacial geology
Glacial features
Pfordte, Otto F., 1905

Geological exploration
Gibbon, Lardner, 1854

Petroleum

Dominican Republic
Siguel, F., 1871

Geologic occurrence, Geographic distribution
Peckham, S. F., 1884

History
Hunt, T. Sterry, 1862
Peckham, S. F., 1884

New Mexico
Herrick, C. L., 1901

Origin
Peckham, S. F., 1884
U. S. Department of Commerce and Labor, 1905

Resources
Hays, S. S., 1866

Technology
Peckham, S. F., 1884

United States
History, Resources
Hays, S. S., 1866

Uses
Peckham, S. F., 1884

Petrology

Daubrée, G. A., 1862

Arctic
Emerson, Benjamin K., 1879

Bibliography
1887-1888
Merrill, George P., 1890

Colorado
Conkling, A. R., 1877c

Great Plains
Hitchock, Edward, 1853
Shepard, Charles U., 1853

History
Teall, J. J., Harris, 1903
1887-1888
Merrill, George P., 1890

Lake Superior region
Burt, William A., & Hubbard, Bela, 1849
Locke, John, 1849

Michigan
   Burt, William A., & Hubbard, Bela, 1849
New Mexico
   Conkling, A. R., 1877c
Nicaragua
   Endlich, Frederic M., 1874
Northwest Territories
   Emerson, Benjamin K., 1879
Oklahoma, Texas
   Hitchock, Edward, 1853
   Shepard, Charles U., 1853
Rock description
   Merrill, George P., 1889
   Merrill, George P., 1891

Philippines
   Areal geology
      Becker, George F., 1901
      Eveland, Arthur J., 1907
      Ickis, E. M., & Field, E. M., 1905
      McCaskey, H. D., 1907
      Smith, Warren D., 1905b
   Economic geology
      Coal
         Smith, Warren D., 1905a
         Smith, Warren D., 1905b
         U. S. War Department, 1905
         Wigmore, H. L., 1904
         Wigmore, H. L., 1905a
         Wigmore, H. L., 1905b
      Construction materials
         Ickis, E. M., & Field, E. M., 1905
      Mineral resources
         Becker, George F., 1901
         Eveland, Arthur J., 1907
         McCaskey, H. D., 1905a

   McCaskey, H. D., 1905b
   McCaskey, H. D., 1907
   Geological exploration
      McCaskey, H. D., 1907
   Geomorphology
      Landform description
         Eveland, Arthur J., 1907
      Soils
         Dorsey, Clarence W., 1907
         Sanchez, Alfred M., 1905
      Springs
         McCaskey, H. D., 1907
   Surveys
      Geological survey
         Brewer, William H., et al, 1905
      Mining Bureau organization
         McCaskey, H. D., 1905b

Phosphate
   South Carolina
      Shaler, N. S., 1873
   United States, southern
      U. S. Department of Agriculture, 1869a

Physics
   Relation to geology
      Lapworth, Charles, 1904

Physiography. See Geomorphology.

Pisces
   California
      Agassiz, Louis, 1855
   Great Plains, Tertiary
      Leidy, Joseph, 1872b

Pisces (Cont'd.)

Kansas, Cretaceous
Cope, E. D., 1872b

Nebraska
St. John, Orestes H., 1871

New Mexico, Tertiary
Cope, E. D., 1875

Rocky Mountains, Tertiary
Cope, E. D., 1886
Leidy, Joseph, 1872b

Wyoming, Tertiary
Cope, E. D., 1872a
Cope, E. D., 1873
Leidy, Joseph, 1872a

Platinum

History, Resources
Blake, William P., 1869

United States
Pacific slope
Walcott, Charles D., 1906

Portugal

Geomorphology
Geysers, Thermal springs
Peale, A. C., 1883

Surveys
Land and marine surveys
Wheeler, George M., 1885

Porto Rico

Geomorphology
Soils
Dorsey, Clarence W.; Mesmer, Louis, & Caine, Thomas A., 1903

Precambrian

Rocky Mountains
Stratigraphy
Hayden, F. V., 1878c

Pteridospermophyta

United States
White, David, 1905

Q

Quaternary

Arkansas
Little, George, 1882

Erosion, Deposition
Pleistocene
Wallace, A. R., 1894

Europe, fossil man
Suess, Edward, 1873

Germany
Paleoclimatology
Drude, Oscar, 1905

Ireland
Plant distribution
Anderson, Richard J., 1905

Mississippi Valley
Little, George, 1882

North America
Animal and plant distribution
Adams, Charles C., 1905
Plant distribution
Anderson, Richard J., 1905
Harshberger, J. W., 1905
Paleoclimatology
Geikie, James, 1899
Harshberger, J. W., 1905

Quebec
  Geomorphology
    Landform description
      Wilson, A. W. G., 1905
    Glacial geology
      Isostatic rebound
        Bell, Robert, 1898
      Pleistocene lakes
        Coleman, A. P., 1905

R

Railroads. See under Engineering geology.

Reefs
  Florida
    Agassiz, Louis, 1852
    Hunt, E. B., 1864
    Pourtales, L. F., 1871
  Pacific Ocean
    Funafuti atoll
      Sollas, W. J., 1899

Reptilia
  Dinosauria
    Lucas, Frederick A., 1902
  Great Plains, Tertiary
    Leidy, Joseph, 1872b
  Kansas, Cretaceous
    Cope, E. D., 1872b
    Cope, E. D., 1872c
  New Mexico, Tertiary
    Cope, E. D., 1875
  Permian
    Cope, E. D., 1885

Pterodactylia
  Langley, S. P., 1902
Rocky Mountains, Tertiary
  Leidy, Joseph, 1872b
Wyoming, Tertiary
  Cope, E. D., 1873
  Leidy, Joseph, 1872a

Rhode Island
  Geomorphology
    Soils
      Bonsteel, F. E., & Carr, E. P., 1905b
  Marine geology
    Continental shelf
      Lindenkohl, A., 1885

Rivers. See also under Engineering geology; see also Fluvial features under Geomorphology.
  Channel geometry
    Columbia River
      Langfitt, W. C., 1903
    Connecticut River
      Powell, Charles F., 1905
    Delaware River
      Marinden, Henry L., 1882
      Marinden, Henry L., 1883
      Marinden, Henry L., 1885
      Mitchell, Henry, 1881
      Mitchell, Henry 1884
      Mitchell, Henry, 1889a
    Mississippi River
      Bernard, S., & Totten, J. G., 1823
      Mitchell, Henry, 1883
      Warren, G. K., 1878

Rivers (Cont'd.)
Ohio River
Bernard, S., & Totten, J. G., 1823
Deltas
Delaware River
Mitchell, Henry, 1887b
Mississippi River
Delafield, Richard, 1829
Graham, George, 1829
Drainage
San Joaquin River
Soulé, Frank, 1902
Yuba River
Manson, Marsden, 1902
Exploration
Amazon River
Gibbon, Lardner, 1854
Herndon, Lewis, 1853
Amoor River
Collins, P. McD., 1858
Arkansas River
Abert, J. W., 1846
Big Witchita River, Brazos River
Blake, William P., 1856a
Marcy, Randolph B., 1856
Colorado River
Adams, Samuel, 1870
Derby, George H., 1852a
Geikie, Archibald, 1884
Ives, Joseph C., 1858
Ives, Joseph C., 1861
Newberry, J. S., 1861
Parry, C. C., 1869
Powell, John W., 1872a
Powell, John W., 1872b
Powell, John W., 1873
Sitgreaves, L., 1853

Copper River
Abercrombie, W. R., 1900
Allen, Henry T., 1887
Fox River
Warren, G. K., 1876
Koyokuk River
Allen, Henry T., 1887
Mississippi River
Nicollet, I. N., 1843
Owl River
Hoffman, William, 1879
Red River
Marcy, Randolph B., 1853
Ruffner, E. H., 1877a
Shumard, B. F., 1853b
St. Mary's River
Channing, William F., 1848a
Channing, William F., 1848b
Tanana River
Allen, Henry T., 1887
Wisconsin River
Warren, G. K., 1876
Yellowstone River
Barlow, J. W., 1872
Yukon River
Raymond, Charles W., 1871
Zuni River
Sitgreaves, L., 1853
Floods
Mississippi River
Warren, G. K., 1875a
Geologic history
Niagara River
Gilbert, G. K., 1891

Geomorphology

Mississippi River
Long, S. H., & Humphreys, A. A., 1851

Log dams

Big Black River, St. Francis River, White River
Linn, L. F., & Sevier, A. H., 1836

Red River
Fuller, Charles A., 1855
Linnard, T. B., 1844
Long, S. H., 1841
Paxton, Joseph, 1829
Shreve, Henry M., 1833
Shreve, Henry M., & Gratiot, C., 1834

Meanders

Mississippi River
Mitchell, Henry, 1883

Ohio and Mississippi Rivers
Bernard, S., & Totten, J. G., 1823

Sedimentation

Connecticut River
Powell, Charles F., 1905

Mississippi River
Ellet, Charles, Jr., 1851
Ellet, Charles, Jr., 1852
Little, George, 1882
Long, S. H., & Humphreys, A. A., 1851

Stream deflection
Gilbert, G. K., 1885

Surveys

Minnesota River
Warren, G. K., 1875b

Water gaps

Susquehanna River
Howard, William, 1828

Roads. See under **Engineering geology.**

Rocky Mountains. See also names of appropriate states.

Areal geology
Abert, J. W., 1846
Beckwith, E. G., 1855a
Blake, William P., 1855a
Congrès Geologique International, 1893
Cross, Osborne, 1850
Dodge, G. M., 1869
Emmons, S. F., 1893
Emory, William H., 1848
Fremont, John C., 1843
Fremont, John C., 1845
Hall, James, 1852
Hayden, F. V., 1878e
Hayden, F. V., 1878f
Hinton, Richard J., 1890
Lander, Frederick W., 1855
Marcou, Jules, 1855a
Marcou, Jules, 1855c
Stansbury, Howard, 1852
Senate Select Committee, 1838
Stevens, I. I., 1855a
Stevens, I. I., 1855b
Whipple, A. W., 1855

Economic geology

Coal
Hodge, Joseph T., 1872
Lesquereux, L., 1873

Mineral resources
Dodge, G. M., 1868
Dodge, G. M., 1869
Mullan, John, 1863
Peale, A. C., 1872

Rocky Mountains (Cont'd.)

Raymond, Rossiter W., 1869
Raymond, Rossiter W., 1870
Raymond, Rossiter W., 1871
Raymond, Rossiter W., 1872
Raymond, Rossiter W., 1873
Raymond, Rossiter W., 1874
Raymond, Rossiter W., 1875
Raymond, Rossiter W., 1877
Van Lennep, D., 1868

Engineering geology

Railroads

Beckwith, E. G., 1855a
Beckwith, E. G., 1855b
Dodge, G. M., 1868
Dodge, G. M., 1869
Lander, Frederick W., 1855
Schiel, James, 1855a
Stevens, I. I., 1855a
Stevens, I. I., 1855b
Van Lennep, D., 1868
Whipple, A. W., 1855

Railroads, Roads

Mullan, John, 1863

Geological exploration

Abert, J. W., 1846
Adams, Samuel, 1870
Beckwith, E. G., 1855a
Beckwith, E. G., 1855b
Blake, William P., 1855a
Cross, Osborne, 1850
Doane, Gustavus C., 1871
Emory, William H., 1848
Fremont, John C., 1843
Fremont, John C., 1845
Hall, James, 1852
Hayden, F. V., 1872a
Hayden, F. V., 1872b
Hayden, F. V., 1877
Lander, Frederick W., 1855
Marcou, Jules, 1855a
Marcou, Jules, 1855c
Mullan, John, 1863
Powell, John W., 1877
Schiel, James, 1855a
Stansbury, Howard, 1852
Stevens, I. I., 1855a
Stevens, I. I., 1855b
Wheeler, George M., 1874
Wheeler, George M., 1875
Wheeler, George M., 1876
Wheeler, George M., 1877
Wheeler, George M., 1878
Wheeler, George M., 1879a
Whipple, A. W., 1855

Geomorphology

Geysers, Thermal springs

Doane, Gustavus C., 1871

Lacustrine features

Newberry, J. S., 1872

Landform description

Doane, Gustavus., C., 1871
Dodge, G. M., 1869
Powell, John W., 1878b
Thomas, Cyrus, 1872
Van Lennep, D., 1868

Glacial geology

Glacial lakes

Newberry, J. S., 1872

Hydrogeology

Hinton, Richard J., 1890
Hinton, Richard J., 1892
Nettleton, C. E., 1892

Mineralogy

Peale, A. C., 1872

Paleontology
  Invertebrata
    Hall, James, 1852
    Hall, James, 1855
    Meek, F. B., 1872c
    Schiel, James, 1855b
  Paleobotany
    Lesquereux, L., 1872a
  Vertebrata
    Cope, E. D., 1886
    Leidy, Joseph, 1872b
  Stratigraphy
    Hayden, F. V., 1878e
    Hayden, F. V., 1878f
  Cambrian, Precambrian
    Hayden, F. V., 1878c
  Structural geology
    Orogeny
      Hayden, F. V., 1878d
Roumania
  Geomorphology
    Landform description
      de Martonne, E., 1905
  Glacial geology
    Glacial features
      de Martonne, E., 1905
  Surveys
    Land and marine surveys
      Wheeler, George M., 1885
Russia. See Union of Soviet Socialist Republics.

S

Salt
  Brazil
    Herndon, Lewis, 1853

Colombia
  Bureau of the American Republics, 1905a
Oklahoma
  Van Vleet, A. H., 1904
Samoa
  Areal geology
    Steinberger, A. B., 1874
San Domingo
  Surveys
    Land and marine surveys
      Wheeler, George M., 1885
San Salvador
  Structural geology
    Orogeny and uplifts
      Sapper, Karl, 1905
  Volcanology
    Sapper, Karl, 1905
Saskatchewan
  Economic geology
    Gold
      Taylor, James W., 1867
      Taylor, James W., 1868
Sedimentary rocks. See also Sediments.
  Classification
    Merrill, George P., 1891
  Collections
    U. S. National Museum
      Merrill, George P., 1891
  Correlation
    White, Charles A., 1893

Sedimentary rocks (*Cont'd.*)
  Limestone
    California
      Williamson, R. S., 1850
    Mississippi Valley
      U. S. Department of Agriculture, 1870
    Oregon
      Frazer, John F., 1850
    United States, eastern
      U. S. Department of Agriculture, 1869c
    United States, southern, Virginia
      U. S. Department of Agriculture, 1869b
  Physical properties
    Merrill, George P., 1891
  Rate of deposition
    Walcott, Charles D., 1894

Sedimentary structures
  Mudlumps
    Mississippi Delta
      Delafield, Richard, 1829
      Ellet, Charles, Jr., 1851

Sedimentation
  Bay
    Chesapeake Bay
      Hughes, George W., 1837
  Delta
    Mississippi Delta
      Ellet, Charles, Jr., 1851
      Ellet, Charles, Jr., 1852
      Hilgard, Eugene W., & Hopkins, F. V., 1878

  Long, S. H., & Humphreys, A. A., 1851
  Glacial
    Quaternary
      Wallace, A. R., 1894
  River
    Mississippi River
      Ellet, Charles, Jr., 1851
      Ellet, Charles, Jr., 1852
      Long, S. H., & Humphreys, A. A., 1851
    Susquehanna River
      Hughes, George W., 1837

Sediments. See also **Sedimentary rocks.**
  Atlantic Ocean
    Agassiz, Louis, 1872b
    Bailey, J. W., 1856a
    Bailey, J. W., 1856b
    Pourtales, L. F., 1854
    Pourtales, L. F., 1859
    Pourtales, L. F., 1872
  Caribbean Sea
    Agassiz, Alexander, 1881
  Challenger expedition
    Daubrée, A., 1894
  Chesapeake Bay
    Hughes, George W., 1837
  Connecticut River
    Powell, Charles F., 1905
  Florida
    Agassiz, Louis, 1852
    Gibbs, Oliver W., 1856
  Loess
    Little, George, 1882

Marine
  Abyssal
    Daubrée, A., 1894
    Murray, John, 1901
    Murray, John, 1905
  Connecticut, Delaware, Massachusetts, New Jersey, New York, Rhode Island.
    Lindenkohl, A., 1885
  Florida
    Agassiz, Louis, 1852
    Hunt, E. B., 1864
  Marginal
    Daubrée, A., 1894
  Massachusetts
    Mitchell, Henry, 1889b
  Methods
    Craven, T. A., 1855
    Sands, B. F., 1856
    Throwbridge, W. P., 1860

Methods
  Ailuvial harbors
    Mitchell, Henry, 1861
  Marine
    Craven, T. A., 1855
    Sands, B. F., 1856
    Throwbridge, W. P., 1860
  Mississippi Delta
    Forshey, C. G., 1875
    Hilgard, Eugene W., & Hopkins, F. V., 1878
  Mississippi River
    Little, George, 1882
  Origin
    Pilar, George, 1877
  Pacific Ocean
    Clover, Richardson, 1892

Susquehanna River
  Hughes, George W., 1837

Seed ferns, see **Pteridospermophyta.**

Seismology, See also **Earthquakes.**
  Mallet, R., 1860
  Associations
    International Seismological Association
      Gerland, G., 1905
  Bibliography
    1884
      Rockwood, Charles G., Jr., 1885
    1885
      Rockwood, Charles G., Jr., 1886
    1886
      Rockwood, Charles G., Jr., 1889
  History
    1884
      Rockwood, Charles G., Jr., 1885
    1885
      Rockwood, Charles G., Jr., 1886
    1886
      Rockwood, Charles G., Jr., 1889
  Methods
    Palmieri, L., 1871

Shore features. See **Shorelines** below.

Shorelines. See also Shore features under **Marine geology;** Coastal features under **Geomorphology.**
  Changes
    Louisiana
      U. S. Army Corps of Engineers, 1874
    Maine, Masachusetts, Nova Scotia
      Mitchell, Henry, 1880
    Massachusetts
      Boutelle, Charles O., 1887
      Marinden, Henry L., 1890b
      Marinden, Henry L., 1892b
      Marinden, Henry L., 1894
      Marinden, Henry L., 1896
      Mitchell, Henry, 1875
      Mitchell, Henry, 1889b
      Van Ingen, H. S., 1874
      Whiting, Henry L., 1875
      Whiting, Henry L., 1887
      Whiting, Henry L., 1890
    New York
      Bache, A. D., 1856b
  Features
    Delaware, New Jersey
      Mitchell, Henry, 1889a
    Massachusetts
      Marinden, Henry L., 1890a
      Marinden, Henry L., 1892a
      Marinden, Henry L., 1897
      Mitchell, Henry, 1874
      Mitchell, Henry, 1887a
    Nicaragua, Panama
      Mitchell, Henry, 1877
    Oregon, Washington
      Langfitt, W. C., 1903
  Islands
    Gulliver, F. P., 1905

Sediments
  Florida
    Gibbs, Oliver W., 1856
  Tidal features
    Hilgard, J. E., 1875

Siberia. See under **Union of Soviet Socialist Republics.**

Silver
  Andes Mountains, Bolivia, Peru
    Gibbon, Lardner, 1854
  Brazil
    Herndon, Lewis, 1853
  Colorado, Montana, New Mexico, South Dakota, Utah.
    Taylor, James W., 1867
    Taylor, James W., 1868
  History, Resources.
    Blake, William P., 1869
  Lake Superior region
    Jackson, Charles T., 1849
  Nevada
    Blatchly, A., 1867
    Edmunds, J. M., et al, 1863
  United States
    King, Clarence; Emmons, S. F., & Becker, G. F., 1885

Soils
  Analysis
    Hilgard, Eugene W., 1884a
  Chemical analysis
    Philippines
      Sanchez, Alfred M., 1905

Chemistry
Cameron, Frank K., 1900
Cameron, Frank K., 1901

Description

Alabama
Avon-Burke, R. T., et al, 1903
Avon-Burke, R. T., et al, 1904
Bennett, Frank, Jr., & Griffen, A. M., 1904
Bonsteel, F. E., et al, 1907
Carr, E. P., et al, 1907
Hearn, W. Edward, & Geib, W. J., 1908
Jones, Grove B., & Carr, M. E., 1904
McLendon, W. E., & Mann, Charles J., 1907
Smith, Eugene A., 1884b
Smith, William G., & Meeker, F. N., 1905
Smith, William G., & Meeker, F. N., 1907
Wilder, Henry J., & Bennett, Hugh H., 1905

Arizona
Holmes, J. Garnett, 1903b
Holmes, J. Garnett, et al, 1905b
Lapham, Macy H., & Neill, N. P., 1904
Means, Thomas H., 1901

Arkansas
Carter, William T., Jr., et al, 1908
Lapham, J. E., 1903
Loughridge, R. H., 1884b
Martin, J. O., & Carr, E. P., 1904b
Wilder, Henry J., & Shaw, Charles F., 1908

California
Hilgard, Eugene W., 1884b
Holmes, J. Garnett, 1901
Holmes, J. Garnett, 1902
Holmes, J. Garnett, et al, 1904a
Holmes, J. Garnett, et al, 1904b
Holmes, J. Garnett, et al, 1905a
Holmes, J. Garnett, et al, 1905b
Holmes, J. Garnett, & Mesmer, Louis, 1902
Lapham, Macy H., 1904
Lapham, Macy H., & Heileman, W. H., 1902a
Lapham, Macy H., & Heileman, W. H., 1902b
Lapham, Macy H., & Jensen, Charles A., 1905
Lapham, Macy H., & Mackie, W. W., 1905
Lapham, Macy H., & Mackie, W. W., 1907
Means, Thomas H., & Holmes, J. Garnett, 1901
Means, Thomas H., & Holmes, J. Garnett, 1902
Mesmer, Louis, 1904

Colorado
Holmes, J. Garnett, 1904
Holmes, J. Garnett, & Neill, N. P., 1905
Holmes, J. Garnett, & Rice, Thomas D., 1907
Lapham, Macy H., et al, 1903
Means, Thomas H., 1900a

Connecticut
Dorsey, Clarence W., & Bonsteel, Jay A., 1900
Fippin, Elmer O., 1904

Soils (*Cont'd.*)

Delaware
Bonsteel, F. E., & Ayrs, O. L., 1904

Florida
Fippin, Elmer O., & Root, Aldert S., 1904
Griffen, A. M., et al, 1908
Rice, Thomas D., & Geib, W. J., 1905b
Smith, Eugene A., 1884c
Wilder, Henry J., 1907b

Georgia
Avon-Burke, R. T., et al, Marean, Herbert W., 1902a
Carr, M. Earl, & Tharp, W. E., 1908
Ely, Charles W., & Griffen, A. M., 1905
Fippen, Elmer O., & Drake, J. A., 1905a
Lapham, J. E.; Lyman, W. S., & Ely, Charles W., 1907
Marean, Herbert W., 1902
Smith, Eugene A., 1884a
Smith, William G., & Carter, William T., Jr., 1904a

Idaho
Jensen, Charles A., & Olshausen, B. A., 1902
McLendon, W. E., 1904
Mesmer, Louis, 1903

Illinois
Bonsteel, Jay A., et al, 1903a
Bonsteel, Jay A., et al, 1903b
Coffey, George N., et al, 1903a
Coffey, George N., 1903b
Coffey, George N., et al, 1904a
Coffey, George N., et al, 1904b
Coffey, George N., et al, 1904c
Coffey, George N., et al, 1904d
Coffey, George N., et al, 1904e
Fippin, Elmer O., & Drake, J. A., 1905b
Locke, John, 1844
Owen, David D., 1840

Indiana
Avon-Burke, R. T., et al, Ruhlen, La Mott, 1904a
Bennett, Frank, Jr., & Ely, Charles W., 1905
Mangum, A. W., & Neill, N. P., 1905a
Mangum, A. W., & Neill, N. P., 1905b
Marean, Herbert W., 1903a
Neill, N. P., & Tharp, W. E., 1907a
Neill, N. P., & Tharp, W. E., 1907b
Tharp, W. E., & Mann, Charles J., 1908

Iowa
Ely, Charles W.; Coffey, George N., & Griffen, A. M., 1905
Fippin, Elmer O., 1903a
Locke, John, 1844
Marean, Herbert W., & Jones, Grove B., 1904a
Marean, Herbert W., & Jones, Grove B., 1904b
Owen, David D., 1840

Kansas
Burgess, James L., & Coffey, George N., 1905
Burgess, James L.; Tharp, W. E., & Lyman, W. S., 1907

Carter, William T., Jr., & Smith, Howard C., 1908
Drake, J. A., 1904
Drake, J. A., & Tharp, W. E., 1905
Lapham, J. E., & Olshausen, B. A., 1903
Mangum, A. W., & Drake, J. A., 1904

Kentucky
Avon-Burke, R. T., 1904a
Avon-Burke, R. T., 1904b
Griffen, A. M., & Ayrs, Orla L., 1907a
Marean, Herbert W., 1903b
Rice, Thomas D., 1907
Rice, Thomas D., & Geib, W. J., 1905d

Lake Superior region
Jackson, Charles T., 1849

Louisiana
Burgess, James L., et al, 1908
Ely, Charles W.; Marean, Herbert W., & Neill N. P., 1907
Griffen, A. M., & Caine, Thomas A., 1907
Heileman, W. H., & Mesmer, Louis, 1902
Hilgard, Eugene W., 1884c
Jones, Grove B., & Ruhlen, La Mott, 1905
Rice, Thomas D., 1904
Rice, Thomas D., & Griswold, Lewis, 1904a
Rice, Thomas D., & Griswold, Lewis, 1904b

Maryland
Bonsteel, F. E., & Carter, William T., Jr., 1904
Bonsteel, Jay A., 1901a
Bonsteel, Jay A., 1901b
Bonsteel, Jay A., et al, 1902a

Bonsteel, Jay A., & Avon-Burke, R. T., 1901
Dorsey, Clarence W., & Bonsteel, Jay A., 1901
Smith, William G., & Martin, J. O., 1902

Massachusetts
Fippin, Elmer O., 1904

Michigan
Fippin, Elmer O., & Rice, Thomas D., 1902
Geib, W. J., 1908
Hearn, W. Edward, & Griffen, A. M., 1905
Jones, Grove B., & Carr, M. Earl, 1907
McLendon, W. E., & Carr, M. Earl, 1905
Mangum, A. W., & Mann, Charles J., 1905
Rice, Thomas D., & Geib, W. J., 1905c
Wilder, Henry J., & Geib, W. J., 1904b

Minnesota
Bennett, Hugh H., & Hurst, Lewis A., 1908
Caine, Thomas A., & Lyman, W. S., 1905b
Geib, W. J., & Jones, Grove B., 1907
Mangum, A. W., & Schroeder, F. C., 1908
Wilder, Henry J., 1904

Mississippi
Bennett, Frank, Jr., & Winston, R. A., 1908
Burgess, James L., & Tharp, W. E., 1907
Bonsteel, Jay A., et al, 1902b
Caine, Thomas A., & Schroeder, Frank C., 1908

Soils (*Cont'd.*)
  Hearn, W. Edward, & Carr, M. E., 1905
  Hilgard, Eugene W., 1884d
  Martin, J. O., & Ayrs, O. L., 1905
  Smith, William G., & Carter, William T., Jr., 1903
  Smith, William G., & Carter, William T., Jr., 1904b

  Mississippi Valley
  Hilgard, Eugene W., 1884a

  Missouri
  Avon-Burke, R. T., & Ruhlen, La Mott, 1904b
  Carr, M. Earl, & Belden, H. L., 1905
  Drake, J. A., & Strahorn, A. T., 1905
  Fippin, Elmer O., 1903b
  Fippin, Elmer O., & Drake, J. A., 1905b
  Hearn, W. Edward, & Mann, Charles J., 1907a
  Hearn, W. Edward, & Mann, Charles J., 1907b
  Loughridge, R. H., 1884a
  Mann, Charles J., & Tharp, W. E., 1908

  Montana
  Jensen, Charles A., & Neill, N. P., 1903b
  Lapham, Macy H., & Ely, Charles W., 1907

  Nebraska
  Burgess, James L., & Worthen, H. L., 1908
  Hearn, W. Edward, 1904
  Hearn, W. Edward, & Burgess, James L., 1904a
  Kocher, A. E., & Hurst, Lewis A., 1907b
  Martin, J. O., & Sweet, A. T., 1905

  New Hampshire
  Mooney, Charles N.; Westover, H. L., & Bennett, Frank, 1908

  New Jersey
  Avon-Burke, R. T., & Wilder, Henry J., 1903
  Bonsteel, Jay A., & Taylor, F. W., 1902

  New Mexico
  Means, Thomas H., & Gardner, Frank D., 1900

  New York
  Avon-Burke, R. T., & Marean, Herbert W., 1902b
  Bonsteel, F. E.; Carter William T., Jr., & Ayrs, O. L., 1904
  Bonsteel, Jay A., et al, 1904
  Bonsteel, Jay A.; Fippin, Elmer O., & Carter, William T., Jr., 1907
  Carr, M. Earl; Griffin, A. M., & Lee, Ora, Jr., 1908
  Fippin, Elmer O., & Carter, William T., Jr., 1907
  Fippin, Elmer O., & Mann, C. W., 1908
  Hearn, W. Edward, 1903
  Lapham, J. E., & Bennett, Hugh H., 1905
  Mesmer, Louis, & Hearn, W. E., 1903
  Wilder, Henry J., & Belden, H. L., 1905b

  North Carolina
  Caine, Thomas A., 1903
  Caine, Thomas A., & Mangum, A. W., 1903
  Coffey, George N., & Hearn, W. Edward, 1902a
  Coffey, George N., & Hearn, W. Edward, 1902b

Dorsey, Clarence W., et al, 1902
Drake, J. A., & Belden, H. L., 1908
Hearn, W. Edward, & MacNider, G. M., 1908a
Hearn, W. Edward, & MacNider, G. M., 1908b
Kerr, W. C., 1884a
Lapham, J. E., & Lymans, W. S., 1907
Lapham, J. E., & Meeker, F. N., 1904
Mooney, Charles N., & Ayrs, O. L., 1905a
Root, Aldert S., & Hurst, Lewis A., 1907
Smith, William G., 1901
Smith, William G., & Coffey, George N., 1904

North Dakota

Caine, Thomas A., 1904
Caine, Thomas A., & Kocher, A. E., 1904a
Ely, Charles W.; Willard, Rex E., & Weaver, J. T., 1908
Fippin, Elmer O., & Burgess, James L., 1905
Jensen, Charles A., & Neill, N. P., 1903a
Kocher, A. E., & Hurst, Lewis A., 1907a
Rice, Thomas D., 1908

Ohio

Caine, Thomas A., & Lyman, W. S., 1905c
Dorsey, Clarence W., & Coffey, George N., 1901
Lapham, J. E., & Mooney, Charles N., 1907a
Lapham, J. E., & Mooney, Charles A., 1907b
Martin, J. O., & Carr, E. P., 1904a

Meeker, F. N., & Tailby, G. W., Jr., 1908
Rice, Thomas D., & Geib, W. J., 1905a
Smith, William G., 1903a
Smith, William G., 1903b

Oklahoma

Loughridge, R. H., 1884d
McLendon, W. E., & Jones, Grove B., 1908
Rice, Thomas D., & Ayrs, Orla L., 1908

Oregon

Jensen, Charles A., 1904
Jensen, Charles A., & Mackie, W. W., 1904

Pennsylvania

Dorsey, Clarence W., 1901
Martin, J. O., 1904
Smith, William G., & Bennett, Frank, Jr., 1902
Wilder, Henry J., & Belden, H. L., 1905a
Wilder, Henry J., et al, 1907a
Wilder, Henry J.; Strahorn, A. T., & Geib, W. J., 1907

Philippines

Dorsey, Clarence W., 1907
Sanchez, Alfred M., 1905

Porto Rico

Dorsey, Clarence W.; Mesmer, Louis, & Caine, Thomas A., 1903

Rhode Island

Bonsteel, F. E., & Carr, E. P., 1905b

South Carolina

Bennett, Frank, Jr., & Griffen, A. M., 1905
Bonsteel, F. E., & Carr, E. P., 1905a

Soils (*Cont'd.*)

Drake, J. A., & Belden, H. L., 1907a
Drake, J. A., & Belden, H. L., 1907b
Hammond, Harry, 1884
Mangum, A. W., & Root, Aldert S., 1904
Rice, Thomas D., & Taylor, F. W., 1903
Root, Aldert S., & Hurst, Lewis A., 1905
Taylor, F. W., 1903

South Dakota

Bennett, Frank, Jr., 1904

Tennessee

Carr, M. Earl, & Bennett, Frank, 1907
Lapham, J. E., & Miller, M. F., 1902
Lyman, W. S.; Bennett, Frank, & McLendon, W. E., 1908
McLendon, W. E., & Lyman, W. S., 1908
Mooney, Charles N., & Ayrs, O. L., 1905a
Mooney, Charles N., & Ayrs, O. L., 1905b
Safford, James M., 1884
Smith, William G., & Bennett, Hugh H., 1904
Wilder, Henry J., & Geib, W. J., 1904a

Texas

Bennett, Frank, Jr., & Jones, Grove B., 1903
Burgess, James L., & Lyman, W. S., 1907
Caine, Thomas A., & Kocher, A. E., 1904b
Caine, Thomas A., & Lyman, W. S., 1905a
Carter, William T., Jr., & Kocher, A. E., 1905
Carter, William T., Jr., & Kocher, A. E., 1907
Ely, Charles W., & Kocher, A. E., 1908
Hearn, W. Edward, 1904
Hearn, W. Edward, & Burgess, James L., 1904b
Hearn, W. Edward, & Burgess, James L., 1904c
Lapham, J. E., et al, 1903
Lapham, J. E., et al, 1904
Loughridge, R. H., 18884c
Mangum, A. W., & Belden, H. L., 1905
Mangum, A. W., & Carr, M. Earl, 1907
Mangum, A. W., & Lee, Ora, Jr., 1908
Martin, J. O., 1902
Mooney, Charles N., et al, 1907

Utah

Gardner, Frank D., & Jensen, Charles A., 1901a
Gardner, Frank D., & Jensen, Charles A., 1901b
Gardner, Frank D., & Stewart, John, 1900
Jensen, Charles A., & Strahorn, A. T., 1905
Means, Thomas H., 1900b
Sanchez, Alfred M., 1904

Vermont

Wilder, Henry J., & Belden, H. L., 1905b

Virginia

Avon-Burke, R. T., & Root, Aldert S., 1907
Bennett, Frank, Jr., et al, 1908
Bennett, Hugh H., & McLendon, W. E., 1907a

Bennett, Hugh H., & McLendon, W. E., 1907b
Caine, Thomas A., & Bennett, Hugh H., 1905
Carter, William T., Jr., & Lyman, W. S., 1904
Kerr, W. C., 1884b
Lapham, J. E., 1904
Mooney, Charles N.; & Caine, Thomas A., 1902
Mooney, Charles N.; Martin, F. O., & Caine, Thomas A., 1902
Mooney, Charles N., & Bonsteel, F. E., 1903

Washington

Carr, E. P., & Mangum, A. W., 1907a
Carr, E. P., & Mangum, A. W., 1907b
Holmes, J. Garnett, 1903a
Jensen, Charles A., 1902

West Virginia

Caine, Thomas A., & Tailby, G. W., Jr., 1908
Griffen, A. M., & Ayrs, Orla L., 1907b

Wisconsin

Bonsteel, Jay A., 1903
Caine, Thomas A., & Lyman, W. S., 1905b
Geib, W. J., & Jones, Grove B., 1907
Jones, Grove B., & Ayrs, Orla L., 1908
Locke, John, 1844
Meeker, F. N., & Avon-Burke, R. T., 1907
Owen, David D., 1840
Smith, William G., 1904

Wyoming

Neill, N. P., et al, 1904

Evaporation

Briggs, Lyman J., 1900a

Mechanical analysis

Briggs, Lyman J., 1900b

Philippines

Sanchez, Alfred M., 1905

Physical properties

Briggs, Lyman J., 1901

South Africa

Mineralogy

Diamonds

Hatch, F. H., & Corstorphone, G. S., 1906
Williams, Gardner F., 1906

South America

Glacial geology

1903

Holdich, T. H., 1903

Surveys

Land and marine surveys

Wheeler, George M., 1885

South Carolina

Areal geology

Hammond, Harry, 1884
Kieth, Arthur, 1902

Economic geology

Gold

Taylor, James W., 1867

Phosphate

Shaler, N. S., 1873

Geomorphology

Landform description

Hammond, Harry, 1884
Keith, Arthur, 1902

Soils

Hammond, Harry, 1884

South Carolina (Cont'd.)

Soils, Abbeville
Taylor, F. W., 1903

Soils, Campobello
Mangum, A. W., & Root, Aldert S., 1904
Bonsteel, F. E., & Carr, E. P., 1905a

Soils, Cherokee County
Drake, J. A., & Belden, H. L., 1907a

Soils, Darlington
Rice, Thomas D., & Taylor, F. W., 1903

Soils, Lancaster County
Root, Aldert S., & Hurst, Lewis A., 1905

Soils, Orangeburg
Bennett, Frank, Jr., & Griffen, A. M., 1905

Soils, York County
Drake, J. A., & Belden, H. L., 1907b

South Dakota

Areal geology
Coffin, Frederick B., 1892
Culbertson, Thaddeus A., 1851
Culver, Garry E., 1892
Engelmann, Henry, 1858
Jenny, Walter P., 1876
Ludlow, William, 1875
Warren, G. K., 1856
Winchell, N. H., 1874
Winchell, N. H., 1875

Economic geology

Coal, Gold, Silver
Taylor, James W., 1867
Taylor, James W., 1868

Mineral resources
Jenny, Walter P., 1876
Raymond, Rossiter W., 1877

Geological exploration
Culbertson, Thaddeus A., 1851
Engelmann, Henry, 1858
Hayden, F. V., 1872b
Hayden, F. V., 1878b
Jenny, Walter P., 1875
Jenny, Walter P., 1876
Ludlow, William, 1875
Owen, David D., 1852
Powell, John W., 1877
Warren, G. K., 1856
Warren, G. K., 1858
Winchell, N. H., 1874
Winchell, N. H., 1875

Geomorphology

Caves
Gamble, Robert J., 1902

Landform description
Hayden, F. V., 1873
Hoffman, William, 1879
Thomas, Cyrus, 1873

Soils
Coffin, Frederick B., 1892
Culver, Garry E., 1892

Soils, Brookings
Bennett, Frank, Jr., 1904

Hydrogeology
U. S. Department of Agriculture, 1890

Paleontology

Invertebrata
Grinnell, George B., 1874
Grinnell, George B., 1875
Meek, F. B., 1872c
Shumard, B. F., 1858

Spain
>   Economic geology
>   >   Mining law
>   >   >   Raymond, Rossiter W., 1869
>   Surveys
>   >   Geological and topographical surveys
>   >   >   Comstock, C. B., 1876
>   >   Land and marine surveys
>   >   >   Wheeler, George M., 1885

Speleology, see Caves.
Springs. See also Thermal springs; Artesian waters and wells; Hydrogeology; Groundwater.
>   Mineral waters
>   >   Arkansas
>   >   >   Haywood, J. K., 1902
>   >   California
>   >   >   Loew, Oscar, 1876a
>   >   Chile
>   >   >   Gilliss, J. M., 1855
>   >   >   Smith, J. Lawrence, 1855
>   >   Japan
>   >   >   Jones, George, 1856a
>   >   Oregon
>   >   >   Dornbach, L. M., 1855
>   >   Wyoming
>   >   >   Yellowstone National Park
>   >   >   >   Hague, Arnold, 1893a
>   >   >   >   Hague, Arnold, 1893b

Stratigraphy. See also under period terms; see also under appropriate area terms.
>   Paleozoic
>   >   Hall, James, 1851c

Structural geology. See also Deformation.
>   Colombia
>   >   Schott, Arthur, 1861a
>   Lineation
>   >   Hobbs, William H., 1905
>   Rocky Mountains
>   >   Hayden, F. V., 1878d

Sudan
>   Areal geology
>   >   Chevalier, Auguste, 1905

Surveys. See also under appropriate area terms.
>   Arctic, Hall, C. F.
>   >   Final report
>   >   >   Hall, Charles F., 1879
>   Challenger expedition
>   >   Daubrée, A., 1894
>   Coast Survey
>   >   History
>   >   >   Henry, Joseph, 1871
>   Colorado River, Powell
>   >   Final report
>   >   >   Powell, John W., 1872a
>   >   Progress report
>   >   >   Powell, John W., 1872b
>   >   >   Powell, John W., 1873
>   40th parallel, King, C.
>   >   Progress report
>   >   >   King, Clarence, 1871
>   >   >   King, Clarence, 1873
>   >   >   King, Clarence, 1874
>   >   >   King, Clarence, 1875
>   >   >   King, Clarence, 1876
>   >   >   King, Clarence, 1877

Surveys *(Cont'd.)*
King, Clarence, 1878
King, Clarence, 1880

Geological, topographical and marine

Foreign
Comstock, C. B., 1876
Wheeler, George M., 1885

Government land and marine surveys

United States
Wheeler, George M., 1885

History
Baird, Spencer F., 1852
Baird, Spencer F., 1853
Baird, Spencer F., 1855

United States
Hayden, F. V., 1878a
House Committee on Public Lands, 1874
Humphreys, A. A., 1878a
Humphreys, A. A., 1878b
McCrary, George W., 1879
Marsh, O. C., 1878
Marsh, O. C., 1879a
Marsh, O. C., 1879b
Powell, John W., 1878a
Powell, John W., 1879
U. S. War Department, 1874
U. S. War Department, 1878
Wilson, Joseph S., 1868
Wheeler, George M., 1879b
Wright, H. G., 1879
Wright, H. G., 1880

Minesota River

History
Warren, G. K., 1875b

Montana, Hayden, F. V.

Progress report
Hayden, F. V., 1872a

Nebraska, Hayden, F. V.

Final report
Hayden, F. V., 1871

New Jersey Geological Mapping
Kümmel, Henry B., 1905

100th Meridian, Wheeler, G. M.

Progress report
Macomb, M. M., 1882a
Macomb, M. M., 1882b
Macomb, M. M., 1883
Wheeler, George M., 1872a
Wheeler, George M., 1872b
Wheeler, George M., 1873
Wheeler, George M., 1874
Wheeler, George M., 1875
Wheeler, George M., 1876
Wheeler, George M., 1877
Wheeler, George M., 1878
Wheeler, George M., 1879a
Wheeler, George M., 1880
Wheeler, George M., 1884

Pacific Railroad Surveys
Abbot, Henry L., 1855
Agassiz, Louis, 1855
Antisell, Thomas, 1855
Bailey, J. W., 1855a
Bailey, J. W., 1855b
Beckwith, E. G., 1855a
Blake, William P., 1855a
Blake, William P., 1855b
Blake, William P., 1855c
Conrad, T. A., 1855a
Conrad, T. A., 1855b
Conrad, T. A., 1855d
Davis, Jefferson, 1855
Dornbach, L. M., 1855
Emory, William H., 1855
Gibbs, George, 1855
Hall, James, 1855
Humphreys, A. A., & Warren, G. K., 1855

Lander, Frederick W., 1855
Marcou, Jules, 1855a
Marcou, Jules, 1855b
Marcou, Jules, 1855c
Newberry, J. S., 1855
Parke, John G., 1855a
Parke, John G., 1855b
Pope, John, 1855
Schaeffer, George C., 1855
Schiel, James, 1855a
Schiel, James, 1855b
Stevens, I. I., 1855a
Stevens, I. I., 1855b
U. S. War Department, 1855a
U. S. War Department, 1855b
Whipple, A. W., 1855
Williamson, R. S., 1855

Philippines

Mining Bureau
McCaskey, H. D., 1905b

Geological survey
Brewer, William H., 1905

Rocky Mountains, Powell

Progress report
Powell, John W., 1877

Territories, Hayden, F. V.

Progress report
Hayden, F. V., 1873
Hayden, F. V., 1878a
Hayden, F. V., 1878b
Hayden, F. V., 1883a
Hayden, F. V., 1883b

U. S. Army Engineers

Index of publications
U. S. Army Corps of Engineers, 1903

U. S. Geological Survey

Hydrographic Division
Fuller, Myron L., 1905

Hydrology Division
Hollister, George B., 1905

Topographical work
Walcott, Charles D., 1902

U. S./Mexican Boundary

Survey
Chandler, M. T. W., 1857
Conrad, T. A., 1857
Emory, William H., 1857
Hall, James, 1857
Michler, N., 1857a
Michler, N., 1857b
Parry, C. C., & Schott, Arthur, 1857

United States, western

Organization
House Committee on Public Lands, 1874

Wyoming, Hayden, F. V.

Progress report
Hayden, F. V., 1872b

Sweden

Surveys

Geological and topographical surveys
Comstock, C. B., 1876

Land and marine surveys
Wheeler, George M., 1885

Switzerland

Surveys

Geological and topographical surveys
Comstock, C. B., 1876

Land and marine surveys
Wheeler, George M., 1885

T

Taiwan
  Areal geology
    Perry, M. C., 1856
  Economic geology
    Coal
      Jones, George, 1856c
  Geomorphology
    Geysers, Thermal springs
      Peale, A. C., 1883

Tectonics. See also **Orogeny; Uplifts; Epeirogenesis.**
  Causes
    Sollas, W. J., 1901
  Cycles
    Blytt, A., 1890
  Mountain building
    LeConte, Joseph, 1898
  Mountains
    Classification, Origin
      Rice, William N., 1905
  Ocean Basins
    LeConte, Joseph, 1898

Temperature. See **Temperature** under **Earth.**

Tennessee
  Areal geology
    Keith, Arthur, 1902
    Safford, James M., 1884
  Economic geology
    Iron
      Pumpelly, Raphael, 1886
  Geomorphology
    Fluvial features
      Little, George, 1882
    Landform description
      Keith, Arthur, 1902
      Safford, James M., 1884
    Landform description, Loess
      Little, George, 1882
  Soils
    Safford, James M., 1884
    Soils, Davidson County
      Smith, William G., & Bennett, Hugh H., 1904
    Soils, Grainger County
      McLendon, W. E., & Lyman, W. S., 1908
    Soils, Greeneville
      Mooney, Charles N., & Ayrs, O. L., 1905a
    Soils, Henderson County
      Carr, M. Earl, & Bennett, Frank, 1907
    Soils, Lawrence County
      Mooney, Charles N., & Ayrs, O. L., 1905b
    Soils, Madison County
      Lyman, W. S.; Bennett, Frank, & McLendon, W. E., 1908
    Soils, Montgomery County
      Lapham, J. E., & Miller, M. F., 1902
    Soils, Pikeville
      Wilder, Henry J., & Geib, W. J., 1904a
  Stratigraphy
    Quaternary
      Little, George, 1882

Tertiary
  California
    Paleogeography, Pliocene
      Bowman, Amos, 1873

Texas
  Areal geology
    Beale, Edward F., 1858
    Beale, Edward F., & Engle, F. E., 1860
    Blake, William P., 1855c
    Conrad, T. A., 1857
    Emory, William H., 1857
    Hill, Robert T., 1892
    Hall, James, 1857
    Loughridge, R. H., 1884c
    Marcou, Jules, 1855b
    Marcy, Randolph B., 1853
    Marcy, Randolph B., 1856
    Michler, N., 1857a
    Parry, C. C., & Schott, Arthur, 1857
    Pope, John, 1855
    Simpson, J. H., & Marcy, R. B., 1850
    Shumard, B. F., 1853b
    Wislezenus, A., 1848
  Economic geology
    Coal, Copper, Gypsum
      Marcy, Randolph B., 1853
    Iron
      Johnson, Lawrence C., 1888
  Engineering geology
    Railroads
      Pope, John, 1855
    Wells
      Pope, John, 1858
  Geological exploration
    Beale, Edward F., 1858
    Beale, Edward F., & Engle, F. E., 1860

Blake, William P., 1855c
Blake, William P., 1856a
Emory, William H., 1857
Johnston, J. E., et al. 1850
Marcou, Jules, 1855b
Marcy, Randolph B., 1850
Marcy, Randolph B., 1853
Marcy, Randolph B., 1856
Michler, N., 1857a
Parry, C. C., & Schott, Arthur, 1857
Pope, John, 1855
Ruffner, E. H., 1877a
Ruffner, E. H., 1877b
Simpson, J. H., 1850a
Simpson, J. H., & Marcy, R. B., 1850
Shumard, B. F., 1853b
Wheeler, George M., 1877
Wheeler, George M., 1878
Wheeler, George M., 1879a
Wheelock, T. B., 1834
Whiting, William H. C., 1850
Wislezenus, A., 1848

Geomorphology
  Landform description
    Loughridge, R. H., 1884c
    Powell, John W., 1878b
  Soils
    Loughridge, R. H., 1884c
  Soils, Anderson County
    Carter, William T., Jr., & Kocher, A. E., 1905
  Soils, Austin
    Mangum, A. W., & Belden, H. L., 1905
  Soils, Brazoria
    Bennett, Frank, Jr., & Jones, Grove B., 1903
  Soils, Henderson
    Ely, Charles W., & Kocher, A. E., 1908

Texas (Cont'd.)

Soils, Houston County
Carter, William T., Jr., & Kocher, A. E., 1907

Soils, Jacksonville
Hearn, W. Edward, & Burgess, James L., 1904b

Soils, Laredo
Mangum, A. W., & Lee, Ora, Jr., 1908

Soils, Lavaca County
Mooney, Charles N., et al, 1907

Soils, Lee County
Burgess, James L., & Lyman, W. S., 1907

Soils, Lufkin
Hearn, W. Edward, et al, 1904

Soils, Marcos
Mangum, A. W., & Lee, Ora, Jr., 1908

Soils, Montgomery County
Martin, J. O., 1902

Soils, Nacogdoches
Hearn, W. Edward, & Burgess, James L., 1904c

Soils, Paris
Caine, Thomas A., & Kocher, A. E., 1904b

Soils, San Antonio
Caine, Thomas A., & Lyman, W. S., 1905a

Soils, Vernon
Lapham, J. E., et al, 1903

Soils, Waco
Mangum, A. W., & Carr, M. Earl, 1907

Soils, Woodville
Lapham, J. E., et al, 1904

Hydrogeology
Hill, Robert T., 1892
Pope, John, 1858
U. S. Department of Agriculture, 1890

Mineralogy, Petrology
Hitchock, Edward, 1853
Shepard, Charles U., 1853

Paleontology

Invertebrata
Conrad, T. A., 1857
Hall, James, 1857
Ruffner, E. H., 1877b
Shumard, B. F., 1853a

Stratigraphy
Blake, William P., 1856a

Thermal springs

Africa, Australia, California, Chile, China-Tibet, Iceland, Indonesia, Japan, Mexico, Nevada, New Zealand, Portugal, Taiwan.
Peale, A. C., 1883

Arkansas
Haywood, J. K., 1902
Weed, Walter H., 1902

California
Loew, Oscar, 1876a

Colorado
McCauley, C. A. H., 1879
Smart, Charles, 1879

Japan
Jones, George, 1856a

Oregon
Dornbach, L. M., 1855

Wyoming
Barlow, J. W., 1872
Bradley, F. H., 1873
Dana, Edward S., & Grinnell, George B., 1876
Doane, Gustavus, C., 1871
Hague, Arnold, 1893a
Hague, Arnold, 1893b
Hayden, F. V., 1872a
Hayden, F. V., 1883b
Langford, N. P., 1873b
Norris, P. W., 1877
Peale, A. C., 1872
Peale, A. C., 1873
Peale, A. C., 1883

Tibet. See under **China.**

Tin
England, Cornwall and Devon
Hughes, George W., 1844

Trinidad
Economic geology
Mineral resources
Wall, G. P., & Sawkins, Joseph, 1857

Tsunamis
California
1854
Bache, A. D., 1856a
1868
Hilgard, J. E., 1872
Japan
1854
Bache, A. D., 1856a

Turkey
Earthquakes
1881
Rittenhouse, H. O.; Knight, A. M., & Huse, H. P., 1881
Surveys
Land and marine surveys
Wheeler, George M., 1885

U

Union of Soviet Socialist Republics
Areal geology
Siberia
Collins, P. McD., 1858
Economic geology
Metals, Siberia
Palmer, Aaron H., 1848
Geological exploration
Siberia
Collins, P., McD., 1858
Paleontology
Mammalia, Quaternary
Herz, O. F., 1904
Surveys
Geological and topographical surveys
Comstock, C. B., 1876
Land and marine surveys
Wheeler, George M., 1885

United States. See also names of individual states.
Areal geology
Congrès Geologique International, 1893

United States (Cont'd.)

Emmons, S. F., 1893
Hitchock, C. H., 1874
Hitchock, C. H., and Blake, W. P., 1874
Whitney, J. D., 1874

Eastern
Featherstonhaugh, George W., 1836
Long, S. H., 1827
McGee, W. J., et al, 1893

Rocky Mountains
King, Clarence; Emmons, S. F., & Becker, G. F., 1885

Southeast
Featherstonhaugh, George W., 1835

Southwest
Conrad, T. A., 1857
Emory, William H., 1857
Hall, James, 1857
Parry, C. C., & Schott, Arthur, 1857

Western
Blake, William P., 1855a
Davis, Jefferson, 1855
Humphreys, A. A., & Warren, G. K., 1855
King, Clarence; Emmons, S. F., & Becker, G. F., 1885
Marcou, Jules, 1855a
Marcou, Jules, 1855c
U. S. War Department, 1855a
U. S. War Department, 1855b
Whipple, A. W., 1855

Economic geology

Coal
Hitchock, C. H., 1874
Pumpelly, Raphael, 1886

Coal, western
Hayden, F. V., 1868

Construction materials
Hawes, George W., 1884

Diamonds
Hobbs, William H., 1902

Gold
King, Clarence; Emmons, S. F., & Becker, G. F., 1885

Iron
Pumpelly, Raphael, 1886

Limestone, eastern
U. S. Department of Agriculture, 1869c

Limestone, southern
U. S. Department of Agriculture, 1869a

Mineral resources
Browne, J. R., 1868
Featherstonhaugh, George W., 1833
Pumpelly, Raphael, 1886
U. S. Department of the Interior, 1866

Mineral resources, eastern
Featherstonhaugh, George W., 1836

Petroleum
Hays, S. S., 1866
Peckham, S. F., 1884

Phosphate, southern
U. S. Department of Agriculture, 1869a

Platinum, Pacific slope
Walcott, Charles D., 1906

Silver
King, Clarence; Emmons, S. F., & Becker, G. F., 1885

Engineering geology

Railroads, western
Davis, Jefferson, 1855
Humphreys, A. A., & Warren, G. K., 1855
U. S. War Department, 1855a
U. S. War Department, 1855b
Whipple, A. W., 1855

Roads, eastern
Long, S. H., 1827

General

History of geology
Merrill, George P., 1906

Geological exploration
Featherstonhaugh, George W., 1833

Eastern
Featherstonhaugh, George W., 1836

Southeast
Featherstonhaugh, George W., 1835

Southwest
Emory, William H., 1857
Parry, C. C., & Schott, Arthur, 1857

Western
Blake, William P., 1855a
Davis, Jefferson, 1855
Humphreys, A. A., & Warren, G. K., 1855
Marcou, Jules, 1855a
Marcou, Jules, 1855c
U. S. War Department, 1855a
U. S. War Department, 1855b
U. S. War Department, 1874
Whipple, A. W., 1855

Geomorphology

Landform Description
Whitney, J. D., 1874

Hydrogeology

Hollister, George B., 1905

Eastern
Fuller, Myron L., 1905

Marine geology

Continental shelf
Hilgard, J. B., 1885

Paleontology

Invertebrata, southwest
Hall, James, 1857

Invertebrata, western
Hall, James, 1855

Paleobotany
White, David, 1905

Southwest
Conrad, T. A., 1857

Stratigraphy

Hall, James, 1851c

Carboniferous
Hitchock, C. H., 1874

Surveys. See also under **Surveys.**

Land and marine surveys
Wheeler, George M., 1885

National Museum, history
True, Frederick W., 1898

Western surveys
House Committee on Public Lands, 1874
Marsh, O. C., 1878
Marsh, O. C., 1879a
Marsh, O. C., 1879b

United States (Cont'd.)

Topographic mapping

Western

U. S. War Department, 1855a
U. S. War Department, 1855b

Uplifts. See also **Changes in level; Tectonics; Isostatic rebound** under **Glaciation.**

Causes

Davis, W. M., 1905a

Costa Rica, Guatemala, Honduras, Mexico, Nicaragua, Panama, San Salvador.

Sapper, Karl, 1905

Cycles

Blytt, A., 1890

Mechanisms

LeConte, Joseph, 1898

Utah

Areal geology

Dana, James D., 1882
Emmons, S. F., 1893
Engelmann, Henry, 1859
Geikie, Archibald, 1882
Hall, James, 1852
Hayden, F. V., 1883a
Simpson, J. H., 1859
Stansbury, Howard, 1852

Economic geology

Coal

Lesquereux, L., 1873

Gold, Silver

Taylor, James W., 1867
Taylor, James W., 1868

Mineral resources

Browne, J. R., 1867
Browne, J. R., 1868
Peale, A. C., 1872
Raymond, Rossiter, 1869
Raymond, Rossiter, 1870
Raymond, Rossiter, 1871
Raymond, Rossiter, 1872
Raymond, Rossiter, 1873
Raymond, Rossiter, 1874
Raymond, Rossiter, 1875
Raymond, Rossiter, 1877

Geological exploration

Adams, Samuel, 1870
Bradley, F. H., 1873
Dana, James D., 1882
Engelmann, Henry, 1859
Geikie, Archibald, 1882
Hall, James, 1852
Hayden, F. V., 1872a
Hayden, F. V., 1872b
Hayden, F. V., 1873
Hayden, F. V., 1877
Hayden, F. V., 1883a
Jones, William A., 1872
King, Clarence, 1871
King, Clarence, 1873
King, Clarence, 1874
King, Clarence, 1876
King, Clarence, 1877
Parry, C. C., 1869
Peale, A. C., 1873
Powell, John W., 1872a
Powell, John W., 1872b
Powell, John W., 1873
Powell, John W., 1877
Simpson, J. H., 1859
Stansbury, Howard, 1852
Wheeler, George M., 1872b
Wheeler, George M., 1873
Wheeler, George M., 1874
Wheeler, George M., 1875
Wheeler, George M., 1876
Wheeler, George M., 1877
Wheeler, George M., 1878
Wheeler, George M., 1879a

Geomorphology
Landform description
Jones, William A., 1872
Powell, John W., 1878b
Thomas, Cyrus, 1872
U. S. Department of Agriculture, 1871
Soils, Bear River
Jensen, Charles A., & Strahorn, A. T., 1905
Soils, Cache County
Means, Thomas H., 1900b
Soils, Provo
Sanchez, Alfred M., 1904
Soils, Salt Lake Valley
Gardner, Frank D., & Stewart, John, 1900
Soils, Sanpete County
Means, Thomas H., 1900b
Soils, Sevier Valley
Gardner, Frank D., & Jensen, Charles A., 1901a
Soils, Utah County
Means, Thomas H., 1900b
Soils, Weber County
Gardner, Frank D., & Jensen, Charles A., 1901b
Hydrogeology
U. S. Department of Agriculture, 1890
Mineralogy
Peale, A. C., 1872
Paleontology
Invertebrata
Hall, James, 1852
Hayden, F. V., 1883a
Meek, F. B., 1872b
Meek, F. B., 1872c
Meek, F. B., 1873
White, Charles A., 1883
Paleobotany
Lesquereux, L., 1872a
Vertebrata
Cope, E. D., 1872d

V

Venezuela
Economic geology
Mineral resources
Goiticoa, N. V., 1904
Geomorphology
Landform description
Goiticoa, N. V., 1904

Vermont
Economic geology
Iron
Pumpelly, Raphael, 1886
Geomorphology
Soils, Vergennes
Wilder, Henry J., & Belden, H. L., 1905b

Vertebrata. See also class names.
California
Agassiz, Louis, 1855
Chile
Wyman, Jeffries, 1855
Colorado, Idaho, Kansas, Montana, Nevada, New Mexico, Oregon, Wyoming, Tertiary.
Cope, E. D., 1884
Dinosauria
Lucas, Frederick A., 1902

Vertebrata (Cont'd.)

Egypt, Tertiary
Andrews, C. W., 1907

Europe, Quaternary
Rütimeyer, L., 1862

Great Plains, Tertiary
Leidy, Joseph, 1872b

Kansas, Cretaceous
Cope, E. D., 1872b
Cope, E. D., 1872c

Nebraska
Leidy, Joseph, 1852
St. John, Orestes H., 1871

New Mexico, Tertiary
Cope, E. D., 1874
Cope, E. D., 1875

Oklahoma, Quaternary
Holmes, William H., 1903a

Permian
Cope, E. D., 1885

Pterodactylia
Langley, S. P., 1902

Restorations
Lucas, Frederick A., 1901a

Rocky Mountains, Tertiary
Cope, E. D., 1886
Leidy, Joseph, 1872b

Utah, Tertiary
Cope, E. D., 1872d

Wyoming, Tertiary
Cope, E. D., 1872a
Cope, E. D., 1873
Leidy, Joseph, 1872a

Virginia

Areal geology
Brown, C. Newton, 1900
Keith, Arthur, 1902
Kerr, W. C., 1884b
McGee, W. J., et al, 1893

Economic geology
Coal
Brown, C. Newton, 1900
Gold
Taylor, James W., 1867
Iron
Pumpelly, Raphael, 1886
Limestone
U. S. Department of Agriculture, 1869b

Geomorphology
Landform description
Keith, Arthur, 1902
Kerr, W. C., 1884b

Soils
Kerr, W. C., 1884b

Soils, Albermarle
Money, Charles N., & Bonsteel, F. E., 1903

Soils, Appomattox County
Caine, Thomas A., & Bennett, Hugh H., 1905

Soils, Bedford
Mooney, Charles N.; Martin, F. O., & Caine, Thomas A., 1902

Soils, Chesterfield County
Bennett, Frank, Jr., et al, 1908

Soils, Hanover County
Bennett, Hugh H., & McLendon, W. E., 1907a

Soils, Leesburg
Carter, William T., Jr., & Lyman, W. S., 1904

Soils, Louisa County
Bennett, Hugh H., & McLendon, W. E., 1907b

Soils, Norfolk
Lapham, J. E., 1904

Soils, Prince Edward
Mooney, Charles N., & Caine, Thomas A., 1902

Soils, Yorktown
Avon-Burke, R. T., & Root, Aldert S., 1907

Volcanism. See also **Volcanology; Volcanoes.**

Central America
Sapper, Karl, 1905

Magnetic field

Variations
Bauer, L. A., 1905

Mechanisms
Heilprin, Angelo, 1905a

West Indies
Hill, Robert T., 1905b

Volcanoes. See also **Volcanology; Volcanism.**

Alaska

Bogoslov volcanoes
Merriam, C. Hart, 1902

Caribbean region
Hill, Robert T., 1905b

Grand Soufriere, Pelee, Saba
Hovey, Edmund O., 1905a

La Soufriere, Pelee
Anderson, Tempest, & Flett, John S., 1903
Russell, Israel C., 1903

Mt. Misery, Soufriere, Statia
Hovey, Edmund O., 1905b

Pelee
Heilprin, Angelo, 1905a
Heilprin, Angelo, 1905b

Costa Rica, Guatemala, Nicaragua
Jones, James O., 1903

Distribution
Lalleman, G., 1905

Iceland
Boehmer, George H., 1886
Stefansson, Jon, 1907

Italy

Vesuvius
Heilprin, Angelo, 1905a
Lacroix, A., 1907

Mexico

Colima
Sartorius, Charles, 1871

Nicaragua
Chamberlain, P. W., 1903
Dutton, C. E., 1902

Oregon

Crater Lake
Diller, J. S., 1898

Volcanology. See also **Volcanoes; Volcanism.**

Bibliography and history
1884
Rockwood, Charles G., Jr., 1885

Volcanology (*Cont'd.*)
1885
Rockwood, Charles G., Jr., 1886
1886
Rockwood, Charles G., Jr., 1889

W

Washington
  Areal geology
    Gibbs, George, 1855
    Lander, Frederick W., 1855
    Stevens, I. I., 1855a
    Stevens, I. I., 1855b
  Economic geology
    Mineral resources
    Browne, J. R., 1867
    Browne, J. R., 1868
    Mullan, John, 1863
  Engineering geology
    Railroads
    Lander, Frederick W., 1855
    Stevens, I. I., 1855a
    Stevens, I. I., 1855b
    Railroads, Roads
    Mullan, John, 1863
  Geological exploration
    Gibbs, George, 1855
    Lander, Frederick W., 1855
    Mullan, John, 1863
    Stevens, I. I., 1855a
    Stevens, I. I., 1855b
    Wheeler, George M., 1878
    Wheeler, George M., 1879a
  Geomorphology
    Fluvial features, Shore features
    Langfitt, W. C., 1903
    Landform description
    Hergesheimer, E., 1883
    Soils, Everett
    Carr, E. P., & Mangum, A. W., 1907a
    Soils, Island County
    Carr, E. P., & Mangum, A. W., 1907b
    Soils, Walla Walla
    Holmes, J. Garnett, 1903a
    Soils, Yakima
    Jensen, Charles A., 1902
  Glacial geology
    Mt. Adams
    Reid, Harry F., 1905a
  Hydrogeology
    Hinton, Richard J., 1892
  Marine geology
    Shore features
    Langfitt, W. C., 1903
  Weathering, See also **Soils.**
    Massive rocks
    Gilbert, G. K., 1905

West Virginia
  Areal geology
    Brown, C. Newton, 1900
  Economic geology
    Coal
    Brown, C. Newton, 1900
    Jamison, J., 1853
  Geomorphology
    Soils, Upshur County
    Griffen, A. M., & Ayrs, Orla L., 1907b

Soils, Wheeling
Caine, Thomas A., & Tailby, G. W., Jr., 1908

Wisconsin
Areal geology
Featherstonhaugh, George W., 1836
Foster, J. W., & Whitney, J. D., 1851
Locke, John, 1844
Nicollet, I. N., 1843
Norwood, J. G., 1848
Owen, David D., 1840
Owen, David D., 1848a
Owen, David D., 1848b

Economic geology
Copper
Owen, David D., 1848a
Copper, Iron, Lead, Zinc
Locke, John, 1844
Owen, David D., 1840
Copper, Lead
Bell, William H., 1844
Iron
Foster, J. W., & Whitney, J. D., 1851
Pumpelly, Raphael, 1886
Mineral resources
Featherstonhaugh, George W., 1836

Engineering geology
Bridges
Warren, G. K., 1878
Rivers
Warren, G. K., 1876

Geological exploration
Featherstonhaugh, George W., 1836
Locke, John, 1844
Nicollet, I. N., 1843
Owen, David D., 1840

Geomorphology
Fluvial features
Warren, G. K., 1876
Warren, G. K., 1878
Glacial features
Owen, David D., 1848b
Lacustrine features
Bixby, W. H.; Beach, Lansing H., & Gaillard, D. D., 1906

Soils
Owen, David D., 1840
Soils, Carlton
Geib, W. J., & Jones, Grove B., 1907
Soils, Janesville
Bonsteel, Jay A., 1903
Soils, Portage County
Meeker, F. N., & Avon-Burke, R. T., 1907
Soils, Racine County
Jones, Grove B., & Ayrs, Orla L., 1908
Soils, Superior
Caine, Thomas A., & Lyman, W. S., 1905b
Soils, Viroqua
Smith, William G., 1904
Terraces
Desor, E., 1851

Glacial geology
Drift, Glacial lakes
Desor, E., 1851

Paleontology
Invertebrata
Hall, James, 1851a

Wisconsin (Cont'd.)
Nicollet, I. N., 1843
Owen, David D., 1848b
Stratigraphy
Hall, James, 1851b

Wyoming
Areal geology
Dana, Edward S., & Grinnell, George B., 1876
Darton, Nelson H., 1906
Emmons, S. F., 1893
Engelmann, Henry, 1858
Engelmann, Henry, 1859
Jones, William A., 1874
Hague, Arnold, 1893a
Hayden, F. V., 1883a
Hayden, F. V., 1883b
Holmes, William H., 1883
Ludlow, William, 1876
Norris, P. W., 1877
St. John, Orestes H., 1883
Simpson, J. H., 1859
Warren, G. K., 1856

Economic geology
Coal
Lesquereux, L., 1873
Mineral resources
Darton, Nelson H., 1906
Peale, A. C., 1872
Raymond, Rossiter W., 1870
Raymond, Rossiter W., 1871
Raymond, Rossiter W., 1872
Raymond, Rossiter W., 1873
Raymond, Rossiter W., 1877

Geological exploration
Bannister, H. M., 1873
Barlow, J. W., 1872
Bradley, F. H., 1873
Comstock, Theodore B., 1874
Doane, Gustavus C., 1871
Engelmann, Henry, 1858
Engelmann, Henry, 1859
Hayden, F. V., 1872a
Hayden, F. V., 1872b
Hayden, F. V., 1873
Hayden, F. V., 1877
Hayden, F. V., 1878b
Hayden, F. V., 1883a
Hayden, F. V., 1883b
Holmes, William H., 1883
Jones, William A., 1874
Peale, A. C., 1873
St. John, Orestes H., 1883
Simpson, J. H., 1859
Warren, G. K., 1856
Warren, G. K., 1858
Wheeler, George M., 1874
Wheeler, George M., 1875
Wheeler, George M., 1876
Wheeler, George M., 1877
Wheeler, George M., 1878
Wheeler, George M., 1879a

Geomorphology
Geysers, Thermal springs
Barlow, J. W., 1872
Bradley, F. H., 1873
Dana, Edward S., & Grinnell, George B., 1876
Doane, Gustavus C., 1871
Emmons, S., F., 1893
Hague, Arnold, 1893a
Hague, Arnold, 1893b
Hayden, F. V., 1872a
Hayden, F. V., 1883b
Heizmann, C. L., 1844
Langford, N. P., 1873b
Norris, P. W., 1877
Peale, A. C., 1872
Peale, A. C., 1873
Peale, A. C., 1883
Weed, Walter H., 1893

Landform description
Dana, Edward S., & Grinnell, George B., 1876
Doane, Gustavus C., 1871
Gannett, E. M., 1883
Hayden, F. V., 1883b
Holmes, William H., 1883
Langford, N. P., 1873b
Ludlow, William, 1876
St. John, Orestes H., 1883
Thomas, Cyrus, 1872

Mineral waters
Heizmann, C. L., 1874

Soils, Laramie
Neill, N. P., et al, 1904

Hydrogeology
U. S. Department of Agriculture, 1890

Mineralogy
Peale, A. C., 1872

Mineral waters
Heizmann, C. L., 1874

Paleontology
  Invertebrata
  Hayden, F. V., 1883a
  Meek, F. B., 1872b
  Meek, F. B., 1872c
  Meek, F. B., 1873
  Shumard, B. F., 1858
  White, Charles A., 1883

  Paleobotany
  Lesquereux, L., 1872a

  Vertebrata
  Cope, E. D., 1872a
  Cope, E. D., 1873
  Cope, E. D., 1884
  Leidy, Joseph, 1872a

Y

Yukon

Engineering geology
  Railroads
  Powell, John W., 1886b

Z

Zinc

Iowa, Wisconsin
Locke, John, 1844
Owen, David D., 1840

WITHDRAWN

FOR USE IN LIBRARY ONLY